Life at the Gate

Rama Krishna Dasu

Copyright © 2024

All Rights Reserved

Dedication

To my beloved parents, for your endless love and support, and for teaching me the value of perseverance.

To my loving wife and wonderful children,

thank you for your unwavering support and understanding during the time I spent away from you to write this book. Your sacrifices and encouragement mean the world to me.

This journey would not have been possible without your love and patience as I gave up many moments of entertainment to pursue this passion.

I dedicate this book to you, with all my love.

To my teachers and mentors, Your guidance and wisdom have shaped my journey in ways I cannot express. Thank you for igniting my passion for learning, challenging me to grow, and believing in my potential.

To my siblings, for your unwavering belief in me and for always being my safe haven.

To my friends, for the laughter, encouragement, and unforgettable moments that inspire me every day.

This book is for all of you.

Acknowledgments

I want to express my deepest gratitude to everyone who made this book possible. First and foremost, to my wife, Sameera, who has been an integral part of my journey. You have shared with me the realities of life at the gate, and your support has been invaluable. Next, to my family and friends:

Your unwavering support and encouragement have been my foundation. Thank you for believing in me even when I doubted myself. Thank you for your endless patience, laughter, and for always being there to share in my journey. Your feedback and enthusiasm have been invaluable.

To mentors, editors, agents and/or publishers:

It's a great journey with you all throughout this book writing. Your guidance and insights have shaped my writing and helped me complete this book. I am deeply grateful for your expertise and dedication in bringing this book to life. Your hard work has made all the difference.

Lastly, to my readers:

Thank you for your support and for inspiring me to share my stories. I hope this book resonates with you.

With all my appreciation,

Rama Krishna Dasu

About the Author

Rama Krishna is a passionate writer and traveler who believes that every journey holds a story waiting to be discovered. With a master's degree in computer applications, RK brings a unique blend of creativity and analytical thinking to his storytelling.

From an early age, RK developed a love for storytelling, finding joy in weaving tales that captivate and inspire. However, he never seized the opportunity to showcase his writing skills through a book until now. He enjoys spending quality time with family and friends, embracing life with a cheerful and fun-loving spirit.

Currently residing in Mason, Ohio, RK invites readers to embark on this literary adventure and hopes his stories inspire others to explore the world and create their own narratives.

Connect with Rama Krishna at [your website or social media links].

Contents

Dedication i
Acknowledgments ii
About the Author iii
Chapter 1: Arriving in America 1
Chapter 2: Life in India and Towards Marriage 12
Chapter 3: Arranged Marriage 20
Chapter 4: The Wedding 28
Chapter 5: From Delhi to Bahrain 38
Chapter 6: Life in Rome 45
Chapter 7: Mayday, Mayday! 55
Chapter 8: A Day in Paris 62
Chapter 9: Return to America 67
Chapter 10: The Aftermath 77

Chapter 1: Arriving in America

Early one morning, as I woke up, the sounds of baggage drops, a security officer calling out lines, and the bustle of people removing shoes and belts to place bags on the scanning machine filled my ears. These familiar noises brought back vivid memories of an important international journey from India to the United States with my newly married wife.

First of all, let me introduce myself. I am an IT professional who took a voyage from India to the US on a company delegation back in 2008. This was my first international travel experience, and I was a mix of nervous and excited. Like many software engineers in India, visiting the US was a dream come true—a chance to explore a new world, save money, and support my family back home. In a year defined by profound global events—from the economic recession to the rise of new independent states and the tragic terrorist attacks in Mumbai—I found my way to the United States. Despite the turbulence of 2008, this marked a major milestone in my life.

Before my departure, my roommate in Chennai, who had traveled a few months earlier, gave me a rundown of what to expect during the international journey. He walked me through the boarding process, security checks, immigration checks, and what to do after landing in the US.

I landed in Tampa at midnight, only to find that my check-in baggage hadn't arrived. After dealing with the airline's lost and found section, I realized my address book was in the missing luggage. Without any addresses, I decided to stay in a hotel for the night and figure things out the next day. Then, to my horror, I discovered my passport was missing—a serious situation for any immigrant in the US. Retracing my steps, I finally found my passport on a counter where I had asked about taxi services.

Relieved but still anxious, I took a taxi to a hotel run by a friendly North Indian. He kindly let me use his phone to call my family and update them. He also helped me book a cab for the next day and suggested I check out of his hotel the following morning as it was quite far from my office.

Before landing on American soil, I completed my master's degree in 2005 and worked in Chennai, Tamil Nadu, India. My company had decided to relocate me to Tampa, Florida, a coastal city that seemed to hold a particular allure for me. Somehow, through some unseen force, I've always been drawn towards coastal cities. Growing up in Vizag, a beachside city in India, the ocean was a constant presence in my life. So, when I arrived in Tampa, it felt like a natural continuation of that connection to the sea.

When I reached Tampa, I stayed with two different people as roommates near our office. There was a lot of moving that followed because of the job that I had taken up.

Life in Tampa was incredible for me. There were other bachelors who had a lot of energy and a significant amount of good vibes to help me feel at home. If I were to do a headcount, I would say we were a group of ten people who would just have a great time visiting Orlando, Miami, and other great places almost every long weekend. In India, it's common to walk to a nearby grocery store or a friend's house or to spend time in community parks playing cricket. Being out on the streets is a part of daily life. In contrast, here in the U.S., outside of cities like New York and Jersey City, it's rare to see people walking around, and having a car feels almost essential. I remember during my first week in the U.S., I went to a local grocery store and bought three tomatoes for $3. That was when I started doing those currency conversions in my head—something most immigrants do at first. Looking back, it's amusing to think about.

The infrastructure of the United States came as a surprise to me. It wasn't a surprise because I expected anything else from one of the most advanced countries in the world. In the second month of my time in the U.S., I had the chance to visit Miami and Key West with my colleagues. The drive from Miami to Key West remains one of the most breathtaking and scenic experiences I've ever had—a journey through miles of stunning ocean views that felt surreal, like driving on a bridge across the sea.

That trip also marked my first encounter with water sports, an experience that turned out to be both thrilling and nerve-wracking. At the time, I had no experience with water activities and couldn't swim. Our group joined a snorkel team and took a ferry far into the ocean. Once we reached a spot, they threw down ropes, inviting us to snorkel or swim. Without thinking much, I stepped into the water with an inflatable snorkel vest, only to realize a few moments later that I was in the middle of the ocean—without knowing how to swim! The vest helped keep me afloat, but I was barely managing, feeling that panic rises with every second. They had instructed us to signal for help if needed, so after about five minutes of sheer survival, I raised my hand and hopped back onto the boat. Looking back, it was a mix of fear and excitement—scary, but a memory I wouldn't trade.

Not long after, I had my first visit to the Orlando theme parks and later to Manhattan. Both places left me overwhelmed, completely in awe of the vast infrastructure and the limitless possibilities the country seemed to offer.

It might just have a lot to do with the family systems in both countries, where families prefer engaging in indoor activities in India, whereas going to Disneyland is the first wish an American child puts in front of their parents. The phenomenal theme parks, the peaceful beaches, and the serene road trips were concepts somewhat alien to me before I stepped foot in the United States. It is an unreal expectation to want around a billion people to follow the rules, but in India, we aren't used to always following rules. For example, in India, when I stand in front of an elevator, people who come after me usually grab my spot, and I don't see anything odd in it before I visit the US. In the United States, if you stand in front of an elevator, someone who comes after you might just offer you the space first. It was a far-fetched dream in India. However, things have changed a lot in India from a decade ago to now.

Over the years, India has developed infrastructure that might even resemble that of America's grandeur. Still, the true difference lies in the culture of the two nations, and when you talk about culture, you talk about food. Even though Indian cuisine and food have entered America and are readily available in the modern and extremely globalized world, it wouldn't be fair to compare it to what is being eaten in India. You might not find authenticity here, but hey, 100 points for the effort!

The difference in culture was something that I had anticipated before walking into the United States. However, there was some comfort that I left behind in India to attain another kind of comfort in the United States. The comfort of family and that dinner table is set just the way you like it, along with some pacifying words from a loved one after a bad day, which distinguishes life in India from life in America. Being able to have a lighthearted chat with your siblings and twin sisters was severely missed. However, the clean environment, the favorable weather, and the progressive pace made me cling to America. I wasn't far from Indians, to be fair; there were plenty around my workplace who belonged to all parts of India, and there was a sense of familiarity.

It is this sense of familiarity that makes the United States special. Everyone will find people to whom they can relate. Everyone will find someone from their home country, speaking their language and giving just a touch of home. Even though English was our primary medium of communication, the feeling of Hindi and other regional languages made things seem familiar. It was interesting because I'm not fluent in Hindi, but only the sound of it made me connect with some people on a level that I might not have connected with back home in India.

There were potlucks, meetings, and plenty of ways to stay engaged. Weekends were often spent playing board games, with carrom being a favorite. Carrom, a tabletop game of Indian origin, involves flicking discs toward the corners of the board. It's a popular game families and friends play at social gatherings, with different rules depending on the region. One of the funniest—and sometimes frustrating—moments is when a player effortlessly flicks a coin that flies through the air and lands perfectly in the corner, like a miracle shot in snooker.

Another amusing memory from my early days in the U.S. was a prank my roommates played on me. Some of them would enjoy beer or wine on the weekends, while I abstained. One evening, one of my roommates pretended to be overly drunk and out of control, even picking up a laptop and charging toward me. He succeeded in scaring me, but when I realized it was just a prank, I couldn't help but laugh at the whole situation.

When I lived in Tampa, Florida, things were smooth. It was one of those states with a significant immigrant population, which automatically meant more people from South Asia, particularly from my home country, India. Even though my family was back home, which is perhaps where I left my comfort, I ended up making friends in Tampa who were the closest friends I could get to becoming family. We were ten people, and we stayed in the same apartment complex. These friends were Indians from various parts of India, but the feeling of being away from home connected us better than we could have imagined.

Our apartment complex was a rather good one. It was close to where we worked. So, with the familiarity of these people and their relatability, we became comfortable with the practicalities of the situation. Be it the apartment complex provisions like the gym, the pool, the tennis court, going out to bowl on Saturday nights, or visits to Clearwater Beach, there was a sense of comfort that kept all of us connected and going in the right direction with each other. Also, there were trips to other cities within the United States, and we would just go from one place to the other, with our jackets held comfortably in our hands because where one part of the United States could be sunny and warm, the other would be cold and windy.

Speaking of connections, my connection with India was in no way impacted by this. I was still exceptionally close to my motherland and the place of my origin. You might wonder how I knew I was still close to my motherland, India. Well, I knew I was close because I still loved cricket! Being an Indian and being a cricket lover go hand in hand, to be fair. It is almost as if watching, following, and enjoying cricket is a testament to how much you love your country back in India. So, by this parameter and a few more, I was still deeply rooted in India.

I still remember the time during the 2011 Cricket World Cup final when I took leave from work to stay home and watch cricket. Of course, the frequency of watching or playing cricket decreased over time because of the lack of cricket infrastructure in the United States, but the love and dedication remained. With time, however, I started enjoying tennis as well. I had the best of both worlds!

And another thing that connected both my worlds? Calling cards! They may seem and sound outdated in today's time of voice calling over the internet. But, like in our times, legends used calling cards to connect to family back home. These cards would give us limited calling time, and talking to a family back home and ensuring enough time was given to everyone was a task. A task, however, that I would happily indulge in at any given time. They weren't free like international calling is today, but they sure had a charm of their own. Being able to make the effort to talk back home just intensified the bond I already had with my family back home.

Back home, I had three siblings other than myself, along with my parents, who missed me but were also immensely happy that I was doing something on my own and for myself. It is universally known that Indians have big families; well, I had one, too. Indians place high regard on their families and cultural values. So, it would be unfair if I did not mention my relatives, both from the paternal and maternal sides, who make up my family. All my cousins and I would meet at least once a year under the pretext of some occasion. Being the youngest, I shared a bond with everyone in my family, and even though I was physically thousands of miles away from them, emotionally and spiritually, we remained connected. My siblings included an older brother and two twin sisters, who were and still are extremely important to me. It wasn't easy to leave them and the comfort of my motherland to rush into the United States, but now, when I look back, I wouldn't have it any other way.

We don't just like having big families; we also like to celebrate and have big family gatherings. Whether it is a religious occasion, a cultural one, or just something as simple as someone's birthday or anniversary in the family, we will always be meeting. I often reminisce about the closeness and the careless fun of the time. There wasn't much to worry about—just the school exams or maybe angering our parents over some little thing.

When I was in my hometown, I used to meet my cousins, aunts, and uncles several times a year. Especially in January every year for one of our main festivals called Sankranthi. It is a four-day festival in the middle of January. On almost three of those festival days, my Mom's siblings (nine families) met at my grandmother's place, which is also in my hometown, some five kilometers from my home. It was an annual feast for me in my childhood, playing with my loved cousins all day long and eating delicious food prepared by my grandmother. We used to play a variety of games like cards, housie, hide & seek, some book games, and more. Our favorite childhood card game is called 'John,' and it is similar to blackjack in America. Housie is also called Tambola in other parts of India. I remember one of the book games we used to play was called 'Name, place, animal, and thing.' All the participants would pick a random letter each time and write four words; the first two categories represented the name of a person and the name of a place. The other two categories represented an animal and a thing. We also used to play carrom and chess, some of my favorite games.

When I moved to the United States, naturally, I missed many of these events with my immediate and extended family. In modern times, we find options like just video-calling and virtually being a part of events. However, back then, even though being able to talk through webcams was a thing, it wasn't all that popularized in India owing to insufficient advancement in internet infrastructure.

One of these events and weddings that I saw on the webcam with a lot of buffering and disruptions was the wedding of my best friend, Uday. I did try to take leave to attend his wedding, but I couldn't. Life and circumstances, I tell you! As I watched his wedding happen over a small screen in front of me, I could also see the food around there. In all fairness, I could smell the food around there.-Let's go back to where we started this part of the chapter: Tampa, Florida, was a great experience for me, but my move from Florida to Connecticut came as a deep drop into discomfort.

In Florida, I was surrounded by many Indian colleagues, which made it feel like I was working in an Indian IT office. However, when I moved to Connecticut, I was surprised to find that I had only two other Indian colleagues. Despite this, over time, I adapted to the office environment and built connections with the people there.

From my comfortable stay in Florida to an entirely new arena in Connecticut, I found myself to be a little lost and lonely. I didn't know whom to ask around for what because I felt none really related to my concerns and problems.

All the Americans around were pleasant people and welcomed me with nothing but goodwill and wide smiles across their faces. I ended up looking for good accommodation online in an area called Bridgeport. On the surface, it looked like a decent place to live, and I agreed to rent an apartment with a roommate.

Much to my displeasure, however, there was a criminal incident around my apartment, in and with my car. I made myself comfortable inside the apartment and fell asleep, only to wake up the next morning and look at my car broken into. The glass was broken and shattered on the seats and seat covers I had just invested in. I stood there, flabbergasted by what had happened, as I saw my GPS missing.

There was theft! On my first night in that area, there was theft! I had partially made my mind up that I wouldn't want to live in a crime-ridden area like this for too long, and what added fuel to this fire was the group of people I worked with. I mentioned to them that whatever had happened, and they informed me that Bridgeport wasn't the safest area to be around. Had I known people around, and had there been a sense of familiarity with people around Connecticut, I would have taken more advice and opinions regarding where I could move. But I didn't, and well, I made a mistake.

After all of this happened and the fiasco left my mind a little bit, I went back and spoke to my roommate about moving out that very day. He agreed, and I moved to a hotel near my office, and within a month, thankfully, I was in a new apartment.

The Connecticut office may have been a small space, but it was always lively, with celebrations like birthday cake cuttings and monthly team lunches keeping the atmosphere vibrant. Two of my senior teammates, both locals, often took me to Indian restaurants. To my surprise, they could handle even spicier Indian food than I could. With sons around my age, they treated me like a kid but connected with me as friends. Despite the seriousness of the project and the critical nature of the production support, I truly enjoyed the work, thanks to their companionship.

Chapter 2: Life in India and Towards Marriage

The United States became my second home, and it still is. However, a piece of my heart and a chunk of my feelings continue to remain in India, with my family and even my extended family, who always kept me close. After my stint in the United States, I went back home to my parents and was received warmly by my mother, who hugged me and cried heavily. It was almost similar to when I hugged her as a child and cried about whatever I felt. It was interestingly amazing how the tables had turned, and with her emotions, mine skyrocketed likewise. Bliss, happiness, and pleasantness may just be understated here if I describe what I felt at that moment. Perhaps this was the best moment for me to be back, and to be fair, I also want to know why it is the one I remember with so much clarity.

Soon after settling back into my hometown, I met all of my uncles, aunts, cousins, and everyone in the family I used to always be with before I went to the United States. It was one big unit I was walking back into, and I couldn't have been happier around that time with those people. It was a great time, and to date, I cherish every second spent there. But the feeling was distinct and inimitable. Of course, when you live outside your country, you make friends, and there is a sense of familiarity after a while, but nothing comes close to the feeling that closeness with your blood gives you.

I didn't plan on returning to the United States when I returned to India. To me, that was a chapter that ended right there because it might have just been someone else's chance in my company to go there and gain experience. I knew I had to be back to work in Chennai, and after a mere couple of weeks in my hometown, I left for Chennai, where my office was located. It might have been my chance to share some of what I learned with the newer people there. Even though my work back in India and the work that I was doing in the United States were the same, the work ethic and the systems differed. It was only fair in this regard that I joined my Chennai office and taught that to my juniors working there.

As my vacation came to an end, I moved to Chennai. Although I had been there before, I hadn't noticed the discomforts that now stood out. In my defense, I had nothing to compare it to at the time, and it certainly felt like an upgrade from my hometown. However, returning after my time in the United States made me realize how much had changed. My old friends and contacts in Chennai had scattered, and now I found myself primarily staying in touch with my family back home and the friends I had made in the U.S.

Additionally, the climate began to bother me, and there were severe concerns that it would also take a toll on my health. It might just be my luck, but when I moved to Chennai, it was the month of May, which is peak summer in the city. Even though there was a summer and a rather warm one in the United States, everything was indoors, and it was rare that the heat would ever get to you.

In Chennai, however, things were different. Most things were outdoors, and they required walking under the sun to get things done. My body couldn't take it. It had just adapted to the United

States climate, and now here I was, asking for it to adapt back to the Indian summer; it was only fair that the poor guy wouldn't be able to take it. So, as a result, I got sick. In fact, at one point, I got so sick that I had to be in bed for a couple of days completely to recover. It was a wreck, to be honest. However, I have always liked Chennai, and I liked being in India because it was my place, and that is where my roots lay.

In spite of that, I felt like Chennai just wasn't made for me. Apart from the harshness of the climate, there was another reason. You know how sometimes, you see a fantastic suit on the mannequin, and it is offered at a great deal with the ideal price tag, but it just isn't for you? That suit for me was Chennai. I would like it, but then again, it wasn't what would have been my place. Most of my friends from Chennai had moved out of Chennai all this time to the United States. Some were relocated within India, while others were sent to different parts of the world. Well, I wasn't just a sick man having trouble with the climate; I was also a little bit of a lonely man there.

Even my best friend, Uday, had moved to the United States. No friends, and no best friend either. It might not have been the best time to come back from the United States. All my friends there would often convince me to move back there so we could all be together, and in all fairness, there was a lot more opportunity and growth there than in India. However, I truly understood where they were coming from and decided that if my work didn't allow me growth or something along the lines of a promotion, I would look for opportunities outside of this workplace and, of course, outside of India as well. During that time, I visited Tirupati, a pilgrimage site in Andhra Pradesh, known for the Sri Venkateswara Temple perched atop one of the seven peaks of the Tirumala Hills. For many Indians, especially those from the south, a visit to Tirupati is considered a sacred duty, and I felt the same way. The experience brought a deep sense of peace to my mind and renewed my confidence that divine guidance would help me find the right path forward.

It was almost as if my stars and my decision aligned. The company I worked for was looking to relocate many of its workers to the United States. Initially, I wasn't quite part of it because I had just come back from there. I watched them apply for their visas, fill out their applications, and make plans about how they would spend their time there, but the embassy had something else in mind. Most of their visas got rejected because their interviews didn't go well. This wasn't an opportunity I created or even tried to cultivate for myself; it just came my way. As I said, the stars and my decision aligned in the most amazing way they could have.

All these rejections for those people were due to the fact that it was their first application. Here, what randomly worked in my favor was the fact that I already had gotten the visa once, and my process was going to be smoother than anyone else's. My company realized this soon, and then I was sent for the visa interview, which, unsurprisingly, I cleared, and there I was on my way to Tampa again! I went to the United States, into the indoor system, with very little walking under the scorching sun and non-existent pollution. It was heavenly in some ways.

One unforgettable bachelor experience that always comes to mind is the night of December 31st when four colleagues and I spontaneously decided to celebrate the New Year's countdown in Miami. One of them mentioned a restaurant in Miami that was known for its grand celebrations and an abundance of food. It was already 8 p.m., and we had less than four hours until midnight. Without hesitation, I said, "Let's go!" and we hit the road immediately. That night, I drove like a man on a mission—no stops, no fuel breaks, no driver shifts—and luckily, traffic was light. We arrived at the restaurant with just five minutes to spare before midnight and managed to catch the countdown. We all cheered with excitement, celebrating our last-minute success.

From Tampa, I would go and visit my friends in Minneapolis and spend time with them, and it was one of these times that my phone rang. It was the name I often dreaded picking up, and just as I lay my eyes on the screen of my phone, I let out a little disappointment. It was my manager. Nobody likes a call from their manager when they are with their friends, and I am no different. I had to pick the phone up regardless because, well, that's what paid me to stay there, and he gave me a shocker. It wasn't the news that I wanted to hear. I was being moved back to Connecticut! Here I was, with my Indian friends from India, people who were with me in India, and I was being told that I had to return to a place with Indians I could count on my fingers.

It was disappointing, to say the least. Even though I resisted the offer, my manager was persistent about telling me that I had to go. Work was calling. It didn't feel all that great, but I went ahead with it and started to make my living in Connecticut again. It is a lovely place, and there are no doubts or questions about that. However, just like Chennai, it wasn't my place or my suit to wear. As I completed my year there, it was time for me to go back to India for three weeks and attend my brother's wedding.

Weddings back home are never a sight to miss. The colors, the music, the rituals, and the beautiful family reunions are an absolute treat.

The house was an absolute riot, in a wonderful way, of course. I had responsibilities regarding the wedding as it was my brother's wedding, and that is just how it is here. Brothers are there for their brothers, and especially relationships matter a great deal. It was the tradition of keeping up with family no matter what. During the wedding, I unexpectedly found myself enjoying a special status as the "abroad return" guest. Word must have gotten around to the vendors—whether it was the wedding planner, florist, or caterer—that I had come back from the U.S. for my brother's big day. They all offered me a bit of extra respect and attention, which added to the unique experience of the celebration.

Another season that seemed to be getting over was my bachelor season. During my three-week visit to India, my father received calls from different families for arranged marriage proposals for me. It is rather common in India for parents to find families and spouses for their children to be married to and into. The process, as simply as I could put it, means that it is Tinder, but your parents play a part in the application, and you only meet your match if they think everything else is all right. In this process, we met a couple of families, but since I was only there for three weeks, I didn't feel like I should be concluding anything and rushing into this.

It was a matter of my entire life and the girl's lifetime. I couldn't decide in three weeks, and my parents agreed. My bachelorhood survived the test of the Indian home visit, and I went back to the United States to do my usual thing. However, a little later, my family and I came across a potential match on a matrimonial website, and things seemed like they could work better for us together. Yes, arranged marriages have also gotten digitalized. That's the interesting thing about India. They preserve their culture and what comes inherently to them, and instead of disregarding new ideas like the digital world, they just amalgamate the two.

Anyway, this girl seemed perfect, and so did her family. I am not saying this because this girl is now my wife and will certainly read the book, but I genuinely felt that at the moment. My parents met her family in India as well as the girl, and they really liked each other. This was the first and biggest green signal for this relationship to move forward. The next step was for me to meet the girl's family and the girl, of course, and even though my father-in-law suggested I come to India to meet them, I couldn't. I had already taken my vacation for that quarter and beyond, so it wouldn't be fair to ask for another one this soon. This was merely weeks after being back.

The solution was that I would go and meet the girl's uncle in New Jersey. This way, someone from their family would have met me, and things could go forward for and between us. Much to my happiness, they did like me back, and things started progressing in the right direction. From that day onward, I started speaking to the girl who was set to be my wife. The first time we talked was through a webcam, and it wasn't just her on the other side of the screen but her parents, grandparents, and many other family members. If the room could fit more, I reckon the neighbors would have joined as well. Jokes aside, the idea of speaking to my would-be wife in front of so many other people. While on the other side, I remained alone, which was a little bit of a challenge.

I didn't know what to say. I might have done well at the visa interview to get to the United States for the second time, but this was an interview I was extremely unprepared for. However, we managed to talk a little bit and at least get our basic questions out of the way, and with that, our families decided it was best to schedule our wedding three months from then. It was a decision that made everyone involved happy, and during these three months, my fiancé and I would speak on the phone every single day for hours.

The familiarity grew, and we started to really like each other as we went deeper, sharing our stories and experiences. The three months, in some ways, felt like a breeze as we got to talk to each other, and things went incredibly smoothly. But, on the other hand, sometimes the three months felt too long because I was really starting to fall for her and wanted to meet her in person and get married to her.

Chapter 3: Arranged Marriage

If one were to look at it, every single little corner of the world has its own way of getting married. It is beautiful and fascinating but also absurd at the same time. The institution is the same, the outcome is the same, and they are all looking to solidify the same institution to keep social cohesion in place. Indians, however, believe in social cohesion a tad bit more strongly than some of the other parts of the world. In many countries, marriages are solemnized with an 'I do.' We don't do that in India. In India, such events are interestingly solemnized with a 'WE ALL DO!'

'We all' includes you, your potential partner, your father and mother, the partner's father and mother, yours and their siblings, yours and their extended family, and, of course, a marriage broker. If you think this list is long, I might add it could have been longer depending on what part of the subcontinent you come from. Where it might sound absurd and rather unique, it certainly isn't bad in any way. No culture is bad; it's just different. In India, marriage isn't between only two people but two families. All of them need to be on board.

The traditions of an Indian marriage cannot be listed down, for immense diversity exists. For instance, a tradition down south in Chennai might be a no-go zone up North in Delhi. However, one of these traditions you would find all across the 29 States, 7 Union Territories, and 1 Billion people is that siblings get married in chronological order. So, it has to be the elder sibling first. In my case, it meant that my elder brother and two elder sisters would be married before me. Astrology clearly didn't approve of me thinking that way, and before I knew it, there was a mention of my marriage. If you don't know, let me make it clear. My fellow Indians back home only abide by two things with all their ability. The first thing is the law of the land, and the second thing is astrology. It is either the government or the stars. My family makes up at least some percentage of the Indian population that believes in astrology as a significant part of their lives and decision-making.

Due to astrological concerns, a *Pundit* (a Hindu Priest) mentioned to my parents that if they start looking for a potential match for me, their youngest one, it will brighten the chances of marriage for my elder siblings. There would be momentum, in their case, he said. As I said, my parents believed in it. Hence, the quest to find my bride began right away. The search is a little interesting. Families share what is known as 'bio-data' with either the middlemen, the marriage brokers, or an internal matrimony bureau. After that, people interested in your biodata get in touch with you after some help from the mediator. If all goes well, you end up with a partner for life.

The biodata includes some really fascinating details. For example, some people have their height preferences set; some want their partner to be a certain professional, some want astrological compatibility, some want a certain skin color, and some want someone from a certain social stature. If I were to talk about the bio-data and the kind of preferences on it, that would be a whole separate book!

My parents put my biodata everywhere they could, and whatever the Pundit had predicted started to happen. My brother got engaged and then married in 2012, and even my sister got engaged the same year. She remained engaged for some time as her partner was going to travel for work, but the goal was attained. So everything was now back in chronological order, just like the rest of my country would like it. Soon after, my father came across my (now) wife's profile on the online portal where he had put my biodata. He was mightily impressed with the profile and decided to send them an invitation request for more details and involvement.

It was a couple of weeks later that my father-in-law saw that invitation and accepted it. Two men from the families decided to meet at a midpoint as our families lived roughly an hour away from each other. Moreover, my father-in-law preferred that the families meet at a neutral venue until things were taken further. Home is personal to him, as it should be.

It just so happened that around the same time, my (now) wife's family was scheduled to attend a wedding near my hometown, and they decided to meet my family at the time. She, her parents, and her grandfathers came to the restaurant. Whereas, from my side, it was just my parents. Now, let me explain the setup a little bit here. In these situations, the families get to know each other more than the girl and the boy. The girl and boy are mostly shyly sitting across each other and making small talk if allowed. Had I been there, I would have definitely made the small talk, but I wasn't.

My mother, on my behalf, asked and answered all the questions, too. There were conversations about her hobbies and my hobbies and interests. And, of course, amidst all of this, the interests of our families. The day they met, I was eagerly waiting for feedback. It wasn't the easiest, to be fair, since I was in a country that functioned in a completely different time zone. Mother's dearest got in touch with me right after they returned, just because I asked to, and they told me that they were impressed with the girl and her family. We didn't get a detailed conversation until the next morning when my mother and father let me know they were getting a positive feeling about this.

Things were going amazingly well. And that is when my father spoke to my father-in-law about my wife and my meeting. The level of *traditionalness* on this account varies across India, too, of course. In the olden days, back in the 1970s, there were families where couples did not meet each other till the night of their wedding. There are also families where the couples have met each other but are never given the opportunity to talk. One thing I find absolutely intriguing about arranged marriages in India is that they have evolved with time while still somehow maintaining the traditional aspect of it.

Let's look at how arranged marriages were in the past. The generation of our great-grandfathers perhaps only got married and then talked to their wives. My grandfather's generation probably allowed them to see their to-be wives before being called husband and wife. My father's generation could maybe share a gentle hello before marrying that person. My generation had it a little better. My father insisted that my partner-to-be and I talk to each other before anything else goes forward. Even though my father-in-law had no qualms, he wanted to be sure through astrological means. Everything came out perfectly, and my father-in-law got in touch with me to ask when I could visit India to meet them and, of course, my potential bride. As much as I wanted to be back to see them, I had just returned from vacation for my brother's wedding, and asking for another vacation right after wouldn't have been the most professional thing to do.

As a result, my father-in-law suggested that I meet his brother (actually, brother-in-law) in New Jersey and that we could take things from there. That seemed like a feasible option to me, and well, it was also one that wasn't threatening to my stable job. So, I took some time out and visited them at their house over the weekend. It was a pleasant experience. I spoke with my fiancée's uncle over the phone and scheduled a meeting. The phone conversation was short, and I felt he was nice. I thought it was a casual meet and had nothing much to do with the final decision from the girl's end. But to my surprise, it was much more than that. I went to my fiancée's uncle's home. To my surprise, the uncle's family and two of his cousins were also there.

The family was well-educated and well-spoken. So, three of my fiancée's uncles spoke to me, and I felt like I was in a panel interview at one point, but we all appreciated the humor in the situation. It was sweet to know that they cared about their niece. We had a nice talk; they were very pleasant and friendly. I also understood the bond they all had here and with their families in India and how well they were connected. The conversations were all casual discussions about my work, hobbies, interests, etc. One of the uncles asked me about my weekend hobbies and whether I visited casinos anytime in the US. I sensed that I had to be cautious, as I didn't know what they might think if I said, 'Yes, I go to casinos with friends occasionally.' I knew all was good as long as you did not gamble. So, I wanted everything to be transparent and honest, so I was genuine with my answers, and it turned out they liked my boldness. Later, I learned that my fiancée's uncle's opinion and feedback were crucial to my wife's parents' decision on the match.

They liked me, I liked them, and, thanks to video-calling options, my wife liked me too. My wife and I talked to each other over the webcam, with her family all around her.

Since my parents met my fiancée in a hotel in Vizag, I met her uncle's family in New Jersey at their home, and I had a video call with my fiancée and her parents online. The next step was the big decision.

All the pieces of this arranged marriage puzzle were coming together. Even with that conversation around entire families around us, my wife and I clicked, and that was a majorly huge and positive sign for me. My wife's family agreed to proceed and pursue the match and soon got in touch with my father to visit them at their house. However, it was time for my father to be in a little bit of a dilemma. Remember how I told you marriages were supposed to happen in the descending order of siblings? Yeah. My sister was engaged but not married at the time. This put my father in two minds, as he thought I was younger and shouldn't be married before my elder sister.

My father-in-law called my father and invited my family to their home, which is 45 km from our home in Vizag; their town name is Anakapalle, and it used to be part of the same district as my parents. It wasn't that my father didn't want to finalize the marriage or the proposal. On the contrary, he was all up for it. It was the ceremony that he wanted to be after my sisters. My father and father-in-law spoke to each other in detail. My father asked for a couple of weeks for confirmation, which my father-in-law happily agreed to. In those couple weeks, my father thought a lot about the situation, and one day, my father-in-law decided to call for an update. Just as my father put that phone down, he made me a call to ask me what I wanted to do. Yes, we get these choices back home; do not be surprised. I told my parents that if they were positive about the girl, then we should proceed because I liked her and her family when I spoke to them over the video call.

"What do you want to do about it?" he asked. "Go ahead," I responded with assurance. Rest, they say, is history!

But my parents said the final decision would come after my parents visited the girl's home. The following week, my parents planned to visit my fiancée's home. My father, mother, brother, sister-in-law, and sisters visited my in-laws. This was more like an official gathering prior to the decision.

During that gathering, my mother asked the girl some questions, and she answered accordingly. Once the decision is made, there will be a ritual called 'Thambulalu' (in my native language), where both families exchange a coconut along with some traditional items.

On that day, when my parents visited their home, my father-in-law saw the interest in my parents' eyes towards this match and suggested conducting the above ritual on the same day right away. My father was in a dilemma about whether to proceed, as he is not a quick decision-maker.

Usually, he takes his time to evaluate all possibilities before making a decision. This decision is a big one that is tied to two lives (my wife and I). My father called me and asked whether I could travel from the US to India within the next one or two weeks to meet the girl and her family directly, and they could also see me.

Following that, we could proceed with no ambiguity. But I told them the challenges involved in traveling all of a sudden. And I did not feel a great need to travel to India to finalize the decision. So, I conveyed my 'yes' again. Immediately, my parents informed my father-in-law and his family about the decision, and they were extremely happy, too. There and then, my father and father-in-law went shopping to buy a saree (traditional clothes) for my fiancee and make other arrangements for the upcoming traditional ceremonies.

My mother-in-law made arrangements for essentials at home. On the same day, they completed the Thambulalu ritual, and I was asked to join the occasion via webcam for some time. With that, my search for a girl came to an end.

Chapter 4: The Wedding

Finally, the long-awaited day had arrived, and I couldn't express how excited I felt as the days got closer to our wedding ceremony. It was a great feeling. My fiancé gave me a temporary phone number and loaded local numbers so we could talk to each other. Indian wedding events begin at least a week in advance. The get-togethers, dance performances, music, and lighting are all part of this extravaganza.

The bright and ceremonious celebration began at my house. The lights, loud music, and the environment were worth everything. In India, marriage is not just two people falling in love and getting married. It's more about family, religion, traditions, and caste. These are considered important for the bride and groom's future. A wedding is considered to be a special day in Indian culture. Each wedding has multiple events, which go on for a week or two, and preparations that start months in advance.

Because I was the groom, I was sent to a spa for a facial, pedicure, and manicure. This was my first ever facial; the beautician assured me that it was a smooth process that would leave my skin glowing.

My fiancé texted me the evening I landed to visit her mother and father. My parents told me to return as soon as possible, so I wouldn't suffer the effects of jet lag. I went to their house in the evening after a good nap in the afternoon. My father-in-law gifted me a ring as a welcome present. I texted my fiancé at night to ask if we could meet the next morning and go to the beach.

She eventually agreed. We both went to a nearby beach in Vizag in the morning after I picked her up on my bike. The beach was a wonderful experience. The relaxing atmosphere took me to another world. I could feel the warm sand next to the water. I could hear the sound of the waves pounding the shore. With my fiancé by my side, the view became even prettier.

It is true when they say that no matter where you are in the world, with your soulmate around you, every place feels delightful. It was so quiet and peaceful there that we spent almost an hour together, talking to each other face to face. Even though I was shy, she was far quieter. Neither of us liked to talk much, but we had a mutual feeling of comfort around each other. I was content with the thought that she would be right beside me for the rest of our lives.

However, a wedding is a big step, so I asked her if she had any apprehensions.

"Everything with you is more magical," she said. "I am incredibly happy that you are a part of my life, and I enjoy every second I spend with you."

We spent the next three days shopping. As a typical Indian parent, my father was a little conservative and became concerned about us spending time together before the wedding. Conservative Indian parents believe that meeting your partner before the wedding calls for bad luck. I told my father I really missed her and needed to discuss important matters regarding the wedding in person. As the wedding day arrived, a beautiful celebration commenced. Breathtaking moments of culture, religion, and unity resulted in the perfect ceremony. In traditional

American weddings, the bride mostly wears white, and the groom wears a simple tuxedo. In India, on the other hand, colorful fabrics and heavy gold jewelry are the preferred choices.

The color of my dress was complementary to that of my fiancé. Indian weddings are all about the details of the ceremony. I was a bit nervous but excited at the same time. I was hoping that everything would turn out exactly how we had planned. It was almost time for the ceremony to start, and I was dressed and ready.

My family's excitement and anticipation were very high, as it was a big day. The bride's family usually organizes the main wedding event at their house. My in-laws decorated their house with flowers and lights.

All my relatives, along with my family, arrived at the venue. The bride's family welcomed them with bouquets, music, and food. Over twelve hundred guests blessed our wedding, 85% of whom were from the bride's family. My father-in-law managed everything quite impressively. While greeting all the guests, he even supervised all the activities in a timely manner, including forming a queue for the stage for relatives to wish the couple and directing the photographer to make sure each of the close relatives and friends was part of the video and pictures.

On my arrival, I was warmly welcomed by my mother-in-law. The music and the drumbeats were very loud. With all the emotions running through my body, my heartbeat synchronized with the drumbeat. This is what is called '***baraat'*** in India. My feet were washed, and I was offered milk and honey, as per Indian tradition.

The wedding site, or a Hindu temple, is where the ceremony was to begin. It's a four-pillared, heightened setup called a **'mandap'** and is considered an essential part of the wedding ceremony. The pillars were covered with fresh and beautiful flowers and garlands of mango and leaves. Pots of water were also decorated, which perfectly attracted the light. Right in front of the mandap were chairs arranged in rows for people to observe the ceremony.

There were several vibrant color palettes. Bright shades of red, orange, blue, and pink are especially common. A statue of the God Ganesha was also displayed at the ceremony site. The décor at the temple mostly comprised floral designs, ornate furniture, statues of beloved deities, candles, bright colors, and much more. Both our families put a lot of effort into the decoration. Every decoration for the day was pre-planned. Witnessing the beautiful and traditional decorations at my wedding was heavenly.

Before the wedding rituals began, the priest invoked the blessings of Lord Ganesh. This is performed to bestow good luck upon the couple and their families. Then was the most long-awaited time. I had been waiting breathlessly for this moment. It was time for me to take a first glance at my fiancé as she officially became my wife. This ceremony is the bride's first appearance. My wife walked down the aisle with her uncles and sisters on either side. As soon as the bright and beautiful cloth known as **'anatarpat'** was removed, I could look at my wife lovingly. She looked gorgeous. I felt incredibly lucky and overjoyed to be a part of this celebration. My father-in-law placed her hand on my right hand to show his acceptance. My mother-in-law poured sacred water over my palm. Our families offered their blessings on the union. It was just like a dream. After that, a sacred fire was lit in the pot that was placed in the middle of the mandap. This holy fire is considered 'the witness' to marriage. My wife and I performed our wedding nuptials, and the priest prayed around the sacred fire. This part of Indian culture is called **'jai mala.'** It's when I exchanged my garland with my wife. A knot was tied between my scarf and my wife's saree.

Then, I joined hands with her and walked seven steps in a clockwise direction. This ritual in India represents each round as a promise and principle. It is demonstrated as an act to seek blessings. The circling (pheray) around the fire has an immense and significant meaning. Each round is a prayer to God; the first round is a prayer for plenty of nourishing food. The second is a prayer for a healthy and prosperous life together. The third round is a prayer for wealth and togetherness between the bride and groom to share happiness and pain. The fourth phera asks God to keep increasing love and respect between the couple. The fifth phera is a request for the strength to bear beautiful, heroic, and noble children. The sixth circle prays for a peaceful life together. During the last phera, the couple prays for eternal togetherness, friendship, and loyalty. I have no words to elaborate on how I felt at that moment.

Before the ring exchange, I gifted my wife a necklace that we call the **'mangal sutra'** in India. We exchanged rings, and like any other traditional wedding worldwide, we expressed our commitment to unconditional love and support for each other. In American weddings, the bride and groom feed each other cake; in a traditional Hindu wedding, they exchange sweets. I had asked the photographer to take our pictures because I knew I wanted to keep those memories forever.

Our families gathered around my wife and me at the mandap for a complete family photoshoot. The happiness on each of their faces was unquestionable. It was then time for the final and official step. Both our families gathered to wish us a healthy life ahead. It is believed that good wishes, love, and wisdom are what help build a family structure. As the priest declared us as a couple, we bowed to the crowd and walked back down the aisle. My friends and relatives sprinkled roses all over us. My wife and I touched the feet of our parents, a tradition Indians call **'ashirwad,'** and hugged them to show respect. In India, out of respect for elders, there is no public show. People don't prefer showing affection publicly out of respect for elders.

The traditions are more stringent. The sendoff is another tradition in India that is carried out just like any other normal wedding. My in-laws were all gathered to ensure a proper sendoff for my wife. It was a unique feeling I had never felt before; I experienced an array of emotions during this part.

Now that almost all of the rituals were complete, the tone of the music completely changed to high energy, encouraging the guests to head to the dance floor. My friends and family had put on a show of their own to show their excitement through a dance performance. Hired dancers and choreographers began to entertain the crowd. The dancing became exuberant, and the environment had a joyful vibe. To wish a couple prosperity, another Indian tradition was tossing money at them as they danced.

My sister-in-law attempted to steal and hide my shoes as I was ready to depart the wedding venue. It's a ritual practiced in India to demonstrate both families' open hearts and acceptance. My friends and I had no idea. We had to negotiate to get my shoes back and move forward out of the mandap after some amusing debates and negotiations. I had the time of my life.

After the wedding ended, close family relatives came to our house for the bride's welcoming ritual. My Mom welcomed my wife and me with a pooja thali at the door; she prayed for us and gave us her blessings. Then, there was a bowl full of rice placed in front of my wife, and she moved it with her foot. My cousins decorated my bedroom with flowers and scented candles everywhere. They also draped a curtain all over the side of the bed. As I said earlier, everyone has many responsibilities during a wedding, and each one of the family members is involved equally.

The groom typically plans the honeymoon for the bride, but I did not want to plan our honeymoon outside the country for many reasons. One was that my wife would have to apply for a visa, which is an extensive process. I only had two weeks to complete the procedure. The other reason was that we wanted to explore the core of India together before going to the U.S. since we were not coming back to India any time soon.

Well, I could not let our early days together be ordinary, so I planned a two-day trip from Delhi, the heart of India, to Agra, the city of love. I did not tell my wife about the honeymoon plan because I wanted to surprise her. She didn't know until we landed in Delhi. From there, we took a train to Agra because my wife loved to travel by train. It was a two-hour journey. I had planned to show her the Taj Mahal, as it is considered 'the symbol of love' all around the world. After all, a trip to India is incomplete if one does not visit the Taj Mahal. It is one of the seven wonders of the world for the right reasons. We reached Agra in the morning. Though many say it is not a place for a newlywed couple to visit, I believe it was created as a romantic gesture.

The Taj Mahal is a funerary mosque built by Shah Jahan in memory of his beloved wife, Mumtaz Mahal. Visiting the Taj Mahal with my wife was the best decision ever. I believe it symbolizes love, as it represents the memory of your loved one staying forever. It teaches us that memories should be framed eternally. The Taj Mahal is famous for its breathtaking view. The architectural style of the Taj Mahal is one of a kind, which truly makes it a masterpiece. Every inch of it is embedded with different types of stones. It has always been my dream to visit the Taj Mahal with my loved ones, and I took my chance to manifest my dream into reality. Well, my wife loved the surprise as well. And I could see how happy she was to see the Taj Mahal with the love of her life.

We went to enjoy the north Indian cuisine later that evening. We reached our 5-star hotel in Delhi, and I booked the honeymoon suite. The memories of that trip will forever be embedded in my heart. I remember my wife telling me that she would treasure the memories of our vacation.

We came back home and attended dinners with our families. After a week, we had to visit Chennai for multiple formalities, like medical tests and visa appointments for my wife. This was a requirement for traveling to the U.S., which is a lengthy and draining process. We mostly stayed back in our hotel room and fulfilled the requirements because we wanted to return home as soon as possible. The weather in Chennai stays extremely hot during the summer. This was another reason we got so tired of staying there.

We had to wait for our plane tickets, which would be sponsored by my company after I informed them about the visa approval. During our entire trip to Chennai, we stayed in a hotel. It was the sporting season in European countries, which made it difficult for my company's travel department to find a proper itinerary. My stress level was at its highest because I was getting pressured by my onsite manager and clients to return as soon as possible. My wife got sick because of the heat waves. I had to take care of my wife and handle my workload, and there was no chance of an extended leave. My manager decided that I would report to the Chennai office for eight hours until I returned to the U.S.

It became a never-ending hustle. My wife was sick, and the workload was intense. I was running between the hotel and my office. It was hard to leave my ill wife alone at the office; she needed me around her. My father-in-law and brother-in-law traveled to Chennai to give us a send-off to the U.S., but when they got to know about the delay and unavailability of flights to the States, they insisted we go back home with them till the flights were confirmed. I discussed my situation with my clients and convinced them to let me work from home for one more week.

Then, my wife and I traveled to our home city with her family. The week went smoothly because we got help around us. Well, my company provided the tickets for the following week. We decided to travel directly from our hometown. The original travel itinerary would have started from Vizag to Delhi, then to Bahrain, and eventually to Rome. The flight from Rome would take us to our destination, New York. The journey would be a total of 34 hours long. I was worried for my wife; it was her first time traveling internationally, but I had to accept this long itinerary due to the pressure from my onsite manager.

I kept feeling like I was making a mistake by choosing to take this flight. It would be a long and taxing journey, but I had no other option. I consoled myself with the thought that my wife and I would be there for each other. I later looked back at this and reminded myself to listen to my instincts in the future.

From this point onward, you will get to read the real reason I chose to write this book. Let me tell you in advance that the travel was not a smooth ride. It still is the most horrific travel experience of my life. This was not the worst that could happen to me, but it was tedious. Our traveling time jumped from 34 to 104 hours; we left India on Saturday morning and reached New York on Wednesday night. You are not wrong; I know it sounds dreadful and hard to believe. This is the main story of this book, which starts right in the next chapter. Flip over to read about the terrifying journey of my life.

Chapter 5: From Delhi to Bahrain

My life's most dreaded and arduous journey had already begun the moment the flight took off. Realizing how chaotic the trip had been from the moment it was planned, I should have considered it a warning from God that we should not continue with it.

Just to catch you up on the obstacles that we were already facing before even leaving for the trip, the original plan was to travel from Chennai to New York, but due to the unavailability of proper itineraries, the plan got changed. Therefore, my wife and I decided to start our trip from our hometown, Vizag, instead of Chennai.

My company had already booked tickets from Delhi to New York via Bahrain and Rome, and I booked a domestic flight from Vizag to Delhi separately. Overall, the given travel time was 36 hours, which was already pretty long, but we had no other option but to avoid one more week of delay, as I didn't have enough time to work from home. Since my wife was traveling abroad for the first time, I was worried about her. Though I was perplexed by various thoughts regarding the journey, I managed to console myself with the thought that my wife would be by my side to support me.

Our first flight was on Sunday morning from Vizag to Delhi. We arrived three hours before the airport, which is a very desi thing to do. If it weren't for my experience in traveling, we probably would've arrived five hours before the flight. Our families were present at the airport to send us off, and my Mom was especially happy to see me go to the U.S. as part of a couple instead of a bachelor, and honestly, I was happy about it, too. I finally had the privilege of creating strong and beautiful memories with my wife that I had always desired. Before leaving, my wife and I took a few pictures to capture the delightful memory. Hence, it was a very wholesome feeling, and I will forever cherish those memories.

The first flight from Vizag to Delhi was smooth since we landed on time. The domestic flight was comparatively easier to manage because I was familiar with the people present there, and we could communicate in our first language, Hindi. I did everything I could to reassure my wife because I could tell she was feeling uneasy.

The next flight was to Bahrain, which would be a long flight. I had already comprehended that it would be a dreadful experience as I started to feel homesick, but then I locked my eyes with my wife for just a second and saw my entire home in her honey-brown eyes. With her eyes, she could comfort me as she made me feel like I was at home.

We had a stopover for eight hours in Delhi, which was fine because it was mentioned in our itinerary. We had loads of bags with us, so it was troublesome to carry them everywhere. Hence, we decided to sit at a restaurant at the airport and enjoy some lunch. The weather was pleasant, and I enjoyed every second spent with my wife. We devoured the delicious North Indian food from a great restaurant at the airport. After lunch, we sat in a lounge and started talking. I had precious time there, and I could tell she was a little shy because her expressions were mostly dull, which was understandable considering that it was her first time on an international trip without her family. She missed them, and so did I. However, I felt a sense of relief to have her by my side.

In a typical Indian marriage, the girl usually lives with her husband and his family and attains his family name after marriage. How brave of a woman to leave her family and live with a new family for the rest of their lives. However, in this case, we both had to leave our families and move to my work location in New York, which is more typical of American culture because, over there, couples live separately from their parents. I wanted to show my love and affection for my wife by kissing her, but in India, it is inappropriate in public. It is mostly because of the stringent traditions and out of respect for the elders. However, we still waited for an opportunity to arise.

It was my first time visiting the Delhi International Airport, and my experience was very memorable, especially because it was with my wife, so we both got to share our first experience together. Delhi airport was considerable and beautiful. It was obviously nothing compared to the airports in the U.S., but what made it more beautiful than the rest was my wife standing beside me.

As the service notified us that it was time for boarding, we instantly got up and walked towards the security check. Further, after getting checked twice, we finally got to the plane. I was very much looking forward to this. I was thrilled about this, as we finally got a chance to bond without interruptions. Although my wife did not express much, the excitement was obvious on her face. The flight mostly consisted of Indian and Gulf folks. It's common for Indian youngsters to travel and work in other countries, particularly Gulf countries. I remember reading this newspaper long ago advertising for electricians, plumbers, turners, and fitters in the Gulf. And I realized that in some cases, if youngsters didn't have any higher education like a master's or Ph.D. or qualifications that would offer them good job opportunities in India, they would accept jobs like these in Gulf countries to financially support themselves and their families.

I told her she could order food and drinks whenever she was ready. I wouldn't have minded alcohol since it was allowed on the flight. My wife already had some idea about it because her uncle's family frequently traveled to India from the U.S.

The journey to Bahrain was according to the schedule; we landed there on time and then waited for four or five hours at the airport. I connected my phone to the airport's internet, emailed my family and friends, and updated them that we had reached Bahrain safely and were heading to our next destination, Rome, Italy. I sent that email to my father, my wife's father, my uncle and her elder uncle, and a couple of my close friends who were in the U.S., and also to my friend's husband, who was supposed to come from Stanford to New York J.F.K. airport. Our last destination would be Connecticut, where I would be working.

The folks I emailed don't know each other, so I intended to keep all recipients in BCC, but due to all the stress and fatigue, I accidentally kept them all in CC, where their email accounts were visible to one another. Anyhow, it was still for the best because now they could communicate with each other regarding my flights.

My mother was particularly relieved to hear from me because she tends to worry about me a lot. We finished our third security check of the trip to Bahrain and boarded the flight to Rome. Furthermore, the flight was full, but there were very few Indian families compared to our previous flight—a total of two or three families, including us. Nonetheless, the flight was still enjoyable. It was nice to observe people of different cultures and races. My wife was amused to notice non-Indian people for the first time in real life. My wife felt anxious and a little scared as the plane took off. The visible fear masked her enthusiasm, so I held her hand tightly to comfort her. I whispered into her ear that everything would be okay, and then she looked at me and smiled in gratitude. I loved watching as the plane's wheels took off from the ground, and everything below got smaller and smaller. I let my wife sit on the window seat so she could admire the beauty of the land and sea from the plane. She gasped as she looked down and saw the greenery of the land, which was subtly invisible due to the cloudy weather.

I had wine for the first time on the flight with my wife's permission. I told her that I had tried alcohol a couple of times but was not a fan of it, so she could stop me if she didn't appreciate me drinking alcohol. I genuinely expected her to stop me because, usually, in most Indian families, wives don't like their husbands drinking alcohol. But to my surprise, my wife was open to it. I was astonished that she didn't restrict me from my choices. I felt blessed to have a supportive and loving wife at that time. We got seats on one side, and when the lights were off for nap time, I saw an open window of an opportunity to kiss her, and I utilized it. It was only a peck because my wife was too nervous that someone would see us, even though there were fewer Indians in our surroundings. Overall, I experienced a whirlwind of emotions, and it was an extraordinary feeling.

Later that night, we had plenty of time to talk and share our recent and old memories. She told me about her experience so far since it was her first international trip. She seemed eager about it, and the look on her face gave me butterflies. I tried to crack a few jokes in between, only to see a slight grin on her face. I was successful a few times, but sometimes my jokes failed. Somehow, those gaffs managed to keep her smiling throughout the flight. We spent together on that plane; it was as if she was meant for me. We had great chemistry and shared the same sense of humor.

She would continue it with something funnier even if my jokes didn't land. We tried not to laugh too loud so we wouldn't wake other people, but I guess I was a little tipsy from the wine, so I screamed once or twice. I was ecstatic to discover her comedic side and how gradually she grew accustomed to being around me. We both enjoyed each other's company and were really looking forward to continuing our adventure and reaching our destination.

I was really eager to show her all the places in the US that I visited, especially Times Square in New York, Niagara Falls, etc. It felt like an honor to show my wife all those new and mesmerizing places she had never seen before. Now I know how Aladdin felt when he showed Jasmine the world on his magic carpet. We still had a lot more to travel before reaching New York, so we tried to get as much rest as possible. I gave my wife my pillow so she could sleep comfortably because I knew how frustrating it could have been to sleep on the seats.

However, I managed to sleep for a little while because, most of the time, I was busy making sure my wife slept peacefully. As it was already a long journey with three flights, we were tired, but it was manageable. And that is how my flight ended, in the best way possible. But we still remained unaware of the difficulties we were to endure ahead of us. The journey turned from a fascinating experience to the most horrific experience of my life.

Chapter 6: Life in Rome

The moment has finally arrived, and now the suspense can come to rest because I will be talking about the events that made this trip so frustrating and dreadful.

We finally got to Rome, and I couldn't wait any longer for the last flight of the trip, landing us straight to New York. According to our itinerary, we had a layover of three to four hours at the airport, after which there would be an 11-hour flight. The airport was quite huge, with countless gates leading to different terminals.

After the layover period had passed, we got ready to board our next flight. We got to the security check and then sat straight on the plane. This is where the hassle originated from. What was weird was that we were waiting for 30 minutes for the plane to take off. However, for 30 minutes, the plane wasn't moving or even circling the runway. After 30 minutes, they finally announced that we needed to get off the plane. I was unaware of what happened, but if I had to assume, I would go with the fact that the plane had been delayed since flight delays are generally common.

All the passengers headed toward the counters to figure out what was happening. The airport officials told us that there was a technical difficulty that needed to be fixed, so they gave us an hour to wait, which was understandable because it's just one hour to fly a safe flight. After one hour, we went to the gate again for updates, and they told us to wait another hour. And just like that, they made us wait for three hours and then concluded that the passengers would have to wait for a day as the "technical" issue was a complicated one. They announced that the flight for the day had been canceled and would take off at the same time the next day. So, it would take us another 22-hour wait to get on the next flight.

When cases like this occur, airport officials usually provide the passengers with a transit visa for the local place and a hotel room if the delay is more than six hours. So, just like every other passenger, I was expecting the same thing to happen.

All of us gathered towards the counter near the gate to address the issue and ask them to provide us with some sort of accommodation. It felt like I was in a fish market; everyone was talking, and I could barely hear what the officials were saying about this matter. Once they silenced all of us, they announced that they could not accommodate any of us or provide us with hotel rooms. This was because there was a sports event happening in the city, so almost all the hotel rooms had already been booked. It exasperated me that they could not help us in any way whatsoever.

We were stuck in the airport for the next 20 hours. We couldn't get out of the airport because we did not have a Schengen visa since our layover was supposed to be for three to four hours, technically making us ineligible for the visa.

We only had our American visa, which could only get us to New York and other states in the US, but unfortunately, not anywhere in Rome. A few other families were in the same boat as my wife and me and did not have local visas.

The locals who lived in Rome but were going to New York on vacation had it easier than us because they had just returned to their respective homes. It was infuriating to see them looking as frustrated as us by this delay.

I suddenly heard screams and noises coming from the counter; I ran towards the counter to see what had happened. After 30 minutes of yelling, the officials finally announced that accommodation could only be provided to Schengen visa holders. Since we didn't have a Schengen visa, it was now official that we would be trapped in the airport for another 20 hours. The only thing they could provide us was food coupons for the restaurants within the airport boundaries.

Four out of twelve of us went to the counter and complained that it was impossible to spend an entire 24 hours there without any sleep; the least they could do was offer us the lounge so we could get some rest. But they refused and said that it was under renovation, so it was closed for the time being. All they could "possibly" do was give us food coupons.

I called my booking agent to talk to him regarding this situation and asked if he could help us out. He told me to contact American Airlines since they were the ones who operated the airline. This incident was back in 2012 when I didn't have access to international calling or WhatsApp, so I called my friend in the US and informed him about what had happened; he called to my cell provider and asked them to enable roaming services on my phone.

Once the roaming services were enabled on my mobile, I called American Airlines, and there was a hotchpotch of information. For 20 or 30 minutes, I was on the call with them, having to deal with forwarded calls and discussions just to hear them say that their counter does not exist at the Rome Airport. No one from American Airlines was available here to help us get a transit visa or anything besides the goddamn food coupons. They just told me to follow the airport's protocol on this.

The terminal had 50 to 60 gates, and we had to stay within those gates, so we were basically just running in circles since we had no way out of this. We fought with the airport officials until they left us to rot there. The people who were in the same boat as us formed a small group and decided to meet at the same spot every three or four hours to get updates from each other.

We were roaming around the terminals with our luggage, looking around the food court to get something to eat. This is where the actual issue started. My wife and I are vegetarians, and the food options there were very limited, with no vegetarian options. Every food item had some sort of meat incorporated into the meal. There were no salads or even plain cheese pizza! It was either pepperoni pizza or pizza that was flavored with some other meat, like chicken. And I know what you all might be wondering: why didn't I just buy the pepperoni pizza, pick out the meat from it, and eat it just like that? However, due to my background and culture, I can't eat any food that the meat has touched or is cooked in the same pot. The only options to eat were donuts and muffins. We had food coupons for lunch, dinner, and breakfast, but we couldn't eat anything that would fulfill us for lunch or dinner. However, we managed to survive on the donuts and muffins throughout the day. It was honestly very shocking that they did not have vegetarian options or even food items that didn't even have meat to begin with, like salads or margarita pizza. I was already feeling frustrated and agitated, and being hungry was the cherry on top. I was trying to keep my cool because I had to take care of my wife, and I didn't want to scare or exhaust her more because of the rage I was feeling.

Since we had nothing better to do, we kept roaming around the terminal, just exploring and lowkey hoping to find a lounge or a place to rest our feet. I mean, the officials had already told us the lounge was under construction, so it didn't make sense to go there just to be disappointed and waste our limited energy.

Some guy at the airport directed us to a lounge. He wasn't very specific about the directions, but we made do with it and walked for 30 minutes or an hour until we finally found a lounge. And to my surprise, this lounge was not under any sort of construction. It was perfectly furnished and clean, with no hint that any recent construction had taken place. To say the least, I was utterly disappointed as to why the airport officials lied to us about this. They had no concern or sense of customer service, and it was bothering me.

It was 10 p.m., and I was hungry; to make matters worse, the people at the counter said that the lounge would close by 11 p.m. What was the point of all this? The other lounges are under construction, and the only lounge that was fully built would be closed by 11 p.m.; was this even an airport? Shouldn't it be open 24 hours? And if not, those people knew our circumstances very well, so couldn't they keep it open for the night? So many thoughts were running through my head and adding to my anger.

The lounge had every type of service, like food, a spa, and internet, but no place to sleep or get any rest. Additionally, they charged visitors 50 to 60 dollars per day for using it. Hence, it didn't make any sense to spend another hour at the lounge because anyone there past 11 p.m. would have to pay the next day's charges, a ridiculous amount for the time spent there.

The only option we had for a place to sleep was on those metal chairs with no cushioning. The tricky part was that my wife had brought her gold jewelry, which was in those bags, with her. I was incredibly anxious about that. Obviously, I knew no one would rob us in the middle of an airport, but I still had to take precautions. Hence, my wife and I decided to take turns sleeping. We unpacked some bags to get out a few clothes to work as a blanket and pillow. The chair was not comfortable in the slightest, so we slept on the floor while one of us sat on that chair to look after the luggage.

I started to feel very sleepy; I had gotten there on a really long flight already and had another long flight ahead after this layover. It was getting unbearable. I only got five minutes of sleep on that metal chair. My wife asked me to wake her up in two or three hours, but she was in a deep sleep, so I did not want to disturb her or wake her up. I let her sleep peacefully while I kept an eye on my luggage.

After a while, my wife woke up herself, and she asked me to sleep, so we swapped our shifts to get some rest. I slept for a couple of hours, and then it was finally morning. We went towards the lounge to wait for it to open again. The lounge opened at exactly 5 a.m., so I paid for the day since the flight was at 12 p.m., so we could spend the rest of our time at the lounge.

We freshened up and got some breakfast. The breakfast, the cereal, the bread, and the muffins were delightful. It was especially delightful since we had been practically starving since the day before. The food was freshly made; it wasn't like the packaged, chemically filled food that is basically just poison in a pack. I even had coffee, which was much needed. However, my wife didn't drink coffee; she just had hot milk and some biscuits. We even found cushioned chairs to rest our heads, stretch our legs, and relax until we waited for our flight. It felt so relieving to finally stretch our legs after two days; it was the ultimate satisfaction. I got so comfortable that I rested my eyes for a little bit, too. It was as if the couch and I were one. I realized how much I missed sleeping on a bed, lying down flat on the mattress with my fluffy pillows. Experiences like this make you value basic things that you normally take for granted.

After an hour or so, my wife went to take a shower while I went to the spa to sit on the massage chair and relax. I would never really use these massage chairs, but after two days of flying, it just needed to happen. And it was totally worth it. The full-body massage touched all my sore muscles, and my body just finally felt free. I sat on that chair for 15 minutes. Even though I wanted to sit there for much longer, I had other important things to do as well. It took a lot of strength and willpower to get my butt off that chair. I went to the browsing center to check my emails, and all my family, friends, and colleagues knew that my flight had been canceled since they knew my itinerary, so that was good because I didn't need to update them about it then. Even though they knew about the cancellation, they were still worried about my wife and me. However, my colleagues, on the other hand, were making jokes about it because they thought I had taken a little detour to Rome to have my honeymoon. It was all fun and laughter with them, but I did update them that I was taking that day's flight and heading to the US. My family was especially worried because it was my wife's first international trip, but I assured them there was nothing to worry about and that I was taking good care of her. I even updated them about the fact that the airport officials could not provide us with accommodation and how it was such a hassle, but we finally got the lounge, so we were relaxing there. My family worries a lot, so I had to give them each and every detail so that they were satisfied enough to feel relieved.

This is not what I wanted my wife's first international trip to be like. I felt regretful about it, but I tried my best to make it better for her. It was all just hitting me after my family asked for updates; I realized that this trip was so different from what my wife had imagined. I know I was suffering from this, but my wife must've been going through so much more. Missing her family and then being stuck in an unfamiliar place for 24 hours is a lot. She was very subtle about her feelings, but I could still understand what she was going through.

It was my turn to shower, so I went towards the bathroom and noticed how rich it looked. It had concrete walls, unlike the wooden ones that are usually seen in houses. Even the doors were very fancy, and the shower room was very spacious as well. It felt like I was in a 5-star hotel. However, my admiration didn't last very long because once I was done with the shower, I reached out to the doorknob to open the door, but I was locked inside. Despite my prior appreciation, I didn't love the bathroom enough to be locked inside it. The worst part was that I didn't even have my phone to call my wife. I started panicking. I shouted and screamed for 15 minutes, hoping that someone could hear me and let me out. I was in there for 30 minutes, and I was wondering why my wife hadn't noticed that I had been gone for 30 minutes. She didn't even ask someone to check inside the men's washroom because she couldn't come there, but she didn't do anything to find me. I kept banging on the door non-stop, just wishing that someone would hear me and open the door. I had a flight to catch, and I was starting to feel very anxious and uneasy. I felt like breaking the door at this point, and I did try, but the lock just wouldn't budge. The concrete walls I so admired were starting to make me panic. The fascination I felt about the bathroom was now starting to frighten me. I was surrounded by four concrete walls with just one way out, and that also had been locked, with the worry of my wife, who was sitting alone waiting for me. It was all happening so fast, and I was at the breaking point. I felt claustrophobic and anxious and frustrated and wrecked. It was a lot to take in 30 minutes.

After five more minutes of panic, someone came to clean the washroom, and he finally heard my screams and helped me out of this. He could not open the door from outside, but we communicated through the door and tried to figure out a way to get me out. But the problem was that we couldn't hear each other speaking on either side of the wall. I had to stick my ear to the door to hear what the person was trying to say. It didn't really matter if he could understand what I was saying as long as he knew that I was locked inside the bathroom and I needed to get out. The cleaner went and brought other people to help me out of this, and for ten minutes, they all tried and tried until finally, the door was open. I came out and saw my wife amongst those people as well, so I felt a little relieved to know that she did notice that I had disappeared. I asked her why she had not noticed that I was gone, and she said she thought I was taking a long shower, considering everything.

The lounge had vegetarian options, so we had cheese pizza for lunch, which was satisfying. Around 10.30 a.m., we reached the gate and met the eight other people who were also staying at the airport with us. It was nice to meet them, as it helped us feel less alone in this situation. By 11 a.m., the boarding started, and we boarded the flight again. I just couldn't wait any longer to reach the US, and I was excited that our flight had boarded and was about to reach. Little did we know what the future had planned for us.

Chapter 7: Mayday, Mayday!

Hope and enthusiasm. That's all I remember feeling as our boarding started. After waiting 24 hours for our flight, I finally felt like I could exhale in relief that we were just one flight away from our destination.

The same plane was getting repaired the previous day, which was not what we expected. After talking to the passengers, we all thought our flight was on another plane, but due to the sporting event in Europe, the players from every team had taken over all the planes and hotel rooms, so there was a lot of chaos.

The plane had finally taken off, so we took off our seatbelts, and I readjusted my position to get some rest until we reached New York. After an hour or something, an announcement was made to fasten our seatbelts due to a sudden emergency. It was frightening; I thought the plane was about to crash or something, and I saw my whole life flash before me. Thankfully, it wasn't going to crash; the plane was just facing some issues, and it was going to have minor shifts, so they asked us to fasten our seatbelts to be on the safe side.

A few minutes later, another announcement was made that there was an engine failure, so the plane was flying on a backup engine, and it was shocking that they did not mention this before. Due to this mishap, they planned an emergency landing in Paris to get the engine "fixed" again. All the passengers went into chaos in the last 10 minutes. People were shouting, screaming, and panicking, and rightfully so. The children started crying, and the air hostess could not handle the situation. I was terrified of the plane's chaotic environment and immediately started praying to God for help. Whatever the situation may be, I always go straight to God and pray to Him for guidance.

They got the clearance from Paris, so we were on the way to land there. Surprisingly enough, 10 minutes before landing, they mentioned that the wheels could not open properly, so the pilot would have to do a belly landing, so that was fun to know.

The entire time, I was just wondering why they were making such major announcements so late. They should have been clear and open to all the passengers about what was happening.

The entire landing was a chaotic process, but I was thankful that at least the plane did not crash, considering the events occurring throughout the trip. I held my wife's hand tightly to make her feel safe and calm. I closed my eyes and started praying to God.

They were following the protocol for a belly landing; hence, several trucks gathered around the runway to reduce the plane's damage and ensure a safer landing for the passengers. A belly landing is not fatal if executed properly, so it was a bit of a risky ride; thankfully, we all managed to survive it. Thank God! The plane landed safely, and we all sighed in relief.

We were told to remain on the airplane, where we stayed for over two hours. To justify this, the airline cited an excuse about us being boarded passengers. Therefore, we could not leave the plane and stay with other passengers at the gate. Since the landing in Paris was so sudden, the flight crew was unprepared to deal with this mishap. We ultimately had to wait at the Paris airport for instructions on how to proceed.

Fatigue, anger, and terror combined into one and created this new feeling that made all the passengers panic and feel confused. The shouting and screaming still haunt me to this day.

They finally updated us with some news. The engine was being repaired due to its failure, and they assured us that it would take 2 to 3 hours to fix the problem. After one hour of waiting on the plane, we finally went down to the terminal and waited for more updates on the situation.

There were two options for the passengers: either wait for 2 to 3 hours and travel on the same plane or wait 24 hours and travel on another flight, which would be paid for by the airline. However, the living situation was not paid for, and the soccer players occupied the hotels, so there was no way to find a place to live for a day.

Everyone was going for the second option because they did not feel secure enough to fly on the same plane that almost crashed. I knew the flight from Paris to New York had no land; it was all ocean, and if the same thing happened again, there would be no place to land the plane. No one wanted to risk their lives, so everyone flew on another plane.

However, my situation was different; we had already spent a day at the terminal at Rome airport, and I anticipated the same thing would happen in Paris as well. They wouldn't be able to provide us with any accommodation or a Schengen visa to get outside the terminal, so we would have to spend 24 hours at the airport again. It was a critical situation to be in, and I didn't know what to do. My wife depended on me to decide because she had never been on an international flight, so she was oblivious to all this.

A feeling of despair crept over me. We had already been delayed for hours, and the thought of spending the night in the airport only added to the frustration. The uncertainty of whether we would be able to leave the airport or not was unbearable. The airline's lack of concern for their passengers left a sour taste in my mouth.

As time ticked by, frustration turned into desperation. The prospect of spending another day in limbo seemed like an unbearable reality. When the announcement came that the flight was canceled, relief washed over us. Finally, we could leave this endless cycle of waiting and uncertainty.

Yet, even in the midst of our joy, doubt lingered in my mind. Would the airline really arrange for our accommodations and another flight? Could we trust them to follow through on their promises? The past hours had shattered our trust, leaving us wary and uncertain of what the future held.

To our surprise, there was an American Airlines counter at the Paris airport, and those of us without Schengen visas—which allow you to visit any of the Schengen States, including France—were whisked away to a corner of the airport, a cramped space where we could barely move. They took our passports and told us to wait, promising us a transit visa and a hotel for the night. But as the minutes turned into hours, our hope began to wane. We were hungry and thirsty, trapped in a foreign country with no way to get help. In the midst of this uncertainty, my wife's act of kindness stood out. She had saved a few muffins from our previous flight, and she distributed them among our fellow passengers who were hungry and anxious. It was a small gesture, but it brought a sense of comfort to the group. They thanked us repeatedly and were grateful for even this small act of kindness.

As we waited for news of our transit visa and hotel, I couldn't help but think about how easily we take things in our lives for granted. We live in a world where we can travel freely, with passports and visas granting us access to new lands. But at that moment, we were stripped of our autonomy and left to rely on the goodwill of strangers. It was a humbling experience that made me grateful for the small blessings in life.

Finally, after what felt like an eternity, the airport officials returned with our stamped passports and the promise of a hotel for the night. It was a relief to leave the cramped space and step into the fresh air. We boarded a shuttle, grateful for the opportunity to rest our weary bodies and escape the chaos of the airport.

However, I couldn't shake off a sense of unease. The situation was dire, and every passing minute heightened our sense of desperation.

But then a glimmer of hope appeared. I remembered my uncle, who lived in the United States but had relatives in Paris. I quickly dialed his sister, my aunt, hoping against hope that she would be able to help us.

To my great relief, my aunt picked up the phone. I explained our situation to her, my voice trembling with anxiety. She listened carefully, her tone soothing and calm. It was a comfort just to hear her voice, to know that we weren't completely alone in this foreign land.

My aunt offered her sympathy and support, telling us that she wished she could help us more directly. Unfortunately, she and her husband were out of the country at that moment and unable to come to our aid.

But even in that moment of disappointment, I was grateful for my aunt's kindness. She offered us her son's contact information, telling us that he lived nearby and would be able to help us if we needed it. It was a small gesture that meant the world to us in our moment of need.

As we settled into the hotel room that the airline had provided for us, I couldn't help but think about the power of timing. If I hadn't remembered my uncle's relatives in Paris, if I hadn't called my aunt at just the right moment, we might have continued to feel uncertain and lost.

We headed straight to baggage claim, only to find one of our suitcases missing. We wanted to report our missing baggage but were told there was not much use complaining in Paris and that we should file a complaint at the lost and found section. It was disheartening to learn that we may never see that bag again, as the airline staff did not seem very hopeful about locating it. It was one of our largest bags, and losing it meant missing out on a significant portion of our belongings. However, it was also one less bag for us to carry, so we chose to see the silver lining.

After all the chaos and doubt, we finally arrived at the hotel in Paris, exhausted and emotionally drained. The feeling of uncertainty and helplessness during those three long hours at the airport, waiting for our passports to be stamped, was unbearable. But as we stepped into our hotel room, the comfort of the bed and the soft pillows welcomed us with open arms.

The hotel room was modest but cozy, and the hot shower was a godsend after a long and stressful day. We had a restless night, partly due to jetlag and partly due to the fear of missing our next flight or encountering any other unexpected challenges.

Chapter 8: A Day in Paris

A popular saying goes, "You never know what you have until it's gone." This rings true in many aspects of life, including the value of something as simple as sleeping in a comfortable bed. After being deprived of proper rest for 60 hours of travel, the significance of a good night's sleep on a bed became crystal clear.

In the South Indian state of Andhra Pradesh lies the renowned temple of Thirupathi. Here, devotees embark on a journey to trek the seven peaks of the Thirumala hills to worship Lord Venkateswara (or Balaji), who sits atop them all. The pilgrimage can last up to 15 to 20 hours of waiting in line to see the idol. However, the moment the idol is finally seen, all the agony of waiting disappears, and the feeling of having attained the ultimate goal in life is experienced.

This same sentiment was felt upon seeing the bed in the hotel room after a long and exhausting journey. Though the hotel was of good quality, the lack of energy made exploring and enjoying the amenities impossible.

Instead, exhaustion took over, and the immediate need for sleep took priority. Waking up at 7 a.m. felt like a luxury, and the comparison to the feeling experienced in Thirupathi was simply a humorous observation.

The previous night, we were provided with a phone number to call in the morning to learn about the schedule and timing of the next flight from Paris to New York that we had to take. The next morning, I spoke with some fellow passengers who were also staying at the same hotel. One person said they were given the 2 p.m. flight, while another said they were given the 12 p.m. flight. I told my wife that if we were given the 2 p.m. flight, we would have around 7 hours to spare, allowing us to visit the Eiffel Tower before heading to the airport.

However, when I called the customer care number, they told us to come to the airport by 10 a.m. to board the 12 p.m. flight. This only gave us 2.5 hours, and I was worried about whether we could still visit the Eiffel Tower and make it to the airport in time. I called my cousin, who is my aunt's son, and asked if it was possible to visit the tower and still catch the 12 p.m. flight if we left within the next 30 minutes.

My cousin warned us that the roads would be congested, and it might take us an hour or more to get from the Eiffel Tower to the airport, which could result in us missing our flight. After experiencing multiple unforeseen scenarios during the past three days of travel, we didn't want to take any chances. Having a solid 5-hour sleep was a bonus amidst all the chaos.

We understood that arriving early at the airport would help move things faster and help us avoid the waiting period after the security check. However, arriving late would delay everything, such as boarding and security checks. Furthermore, if the gate is located far away, it would require a long walk. As a result, we decided to stick to our original plan, get ready, and head to the airport.

The hotel was conveniently located near the airport, taking only 15 minutes to reach. We got ready quickly and headed down to the hotel cafeteria, where a breakfast buffet was available. It was our first time tasting Parisian cuisine, and we enjoyed it. We left the hotel on time and arrived at the airport around 10 a.m.

However, after waiting in the boarding line for 30 minutes, we were informed that our names were not on the 12 p.m. flight. An airport staff member then took us to a different counter to wait. He made a few phone calls and spoke to some people before confirming that we were actually booked on the 2 p.m. flight and not the 12 p.m. flight.

I was confused and frustrated, wondering why the customer care number had given us conflicting information. If we had known earlier that we were on the 2 p.m. flight, we could have visited the Eiffel Tower and arrived on time. However, it was already 11 a.m., and we only had three hours left for our flight. We were disappointed but determined not to miss our flight at any cost.

We woke up early and headed down to the hotel cafeteria for breakfast. The hotel was conveniently located near the airport, so we estimated it would take only 15 minutes to reach there. We enjoyed our first taste of Parisian cuisine and then proceeded to the airport, arriving around 10 am. We had enough time to complete the security check and reached the boarding gate almost an hour before the scheduled departure time.

Charles de Gaulle Airport was an impressive facility, and we walked around near the gates. We visited the duty-free shops and grabbed some coffee while enjoying the relaxing atmosphere. We also met another family there in the same situation as us, without a Schengen visa, and waited together for more than three hours before boarding the flight to New York.

We learned that some passengers had managed to catch the earlier 12 p.m. flight. We were excited to finally board our dream flight to New York, which would take around nine hours. It was a smooth and uneventful journey, with no unexpected announcements or disruptions. Interestingly, almost half of the passengers on the flight were from the missed Rome to the JFK flight.

Upon arriving at JFK Airport, everyone on the flight was overjoyed and clapped in celebration. However, for many passengers who were visiting the US for a short vacation, the technical issues in Rome caused them to lose at least two days of their trip.

After clearing the immigration process, we went to the lost and found section to file a complaint about our missing luggage. However, the employee there seemed confused about our itinerary and asked us many questions about where we had been for the last few days. We explained our travel plans in detail, and eventually, he understood our situation.

He informed us that our case was complex and that it might take up to seven days to track down our missing luggage. If it were not found within seven days, we would be eligible for compensation of $500, as we had booked economy-class tickets.

My friend's husband had planned to pick us up from the airport when we arrived. Originally, he was supposed to pick us up on Monday, but since our flight got delayed by three days, he came to pick us up on Wednesday instead.

Another friend of mine and his wife had also traveled from India and arrived the same day as us. Therefore, he came to pick up all three of us. Throughout our journey, he kept track of our itinerary to ensure he would pick us up whenever we landed. After he picked us up, we went directly to their home. My friend's husband had prepared some delicious Indian food for us, which was a real treat for us after not having any Indian food for four days. Seeing the rice and other Indian dishes ready for us felt like we were in heaven. We enjoyed the food thoroughly and appreciated his thoughtfulness and hospitality.

Chapter 9: Return to America

Hindrances are part and parcel of life. You can make as many plans as you want, but Lady Luck often intervenes, taking matters into her own hands and reminding us of the unpredictable nature of fate. Many often compare the journey of life to a voyage full of unexpected twirls and twisty turns. Some of these events shape our character as people, boost our confidence, and ultimately help us define our true selves.

I embarked on the journey to the United States on a silly note right from the start. Landing in Paris was unfortunate and highly unexpected. Looking back, I can say that this was probably for my own good, but at that point, the stay in Paris was a farce for everyone who was bemused by this sudden happening. The engine failure led to a sudden stop in Paris, and our dream of reaching the United States as soon as possible had to wait.

All in all, I always thought of Paris as a beautiful city—one that emits glamour from every corner. Tranquility and luxury would be the bare minimum a person would expect from a city that boasts one of the seven wonders of the world, the Eiffel Tower. However, for us, checking out of the airport in itself was a hurdle. Right after we got out, it dawned upon us that our suitcases were missing. Perplexed, to say the least, I tried in my full capacity to conjure something that would make this short stay in Paris bearable for me and my wife.

Recalling that, I still remember how my wife, of all people, was so unfamiliar with international flights and the hurdles that come with them, contrary to those who were frequent flyers.

Consequently, the major part of setting things up, even if it was just a day, was done by me.

Fast forward a couple of hours, and we finally got the hotel room. It was a blessing at that point since airport management informed us prior to that it was the inability to accommodate us and whether we possessed a Schengen visa or an American visa. That being said, the proceedings there onwards were not very pleasing.

A glimmer of silver lining was the five-hour sleep that we got in the hotel. At 7:30 a.m., we woke up fresh. At that time, a small matter of concern bothered me—when were we going to onboard the plane to the United States of America?

I remember we were given a phone number for the airline's customer support to learn about our flight rescheduling to the United States. A couple of fellow passengers told me that they were given a time slot for the 2 p.m. flight, while others were given the time for the 12 p.m. flight. I called the representative, and to my disappointment, he told me that I only had two and a half hours to reach the airport for the seats that were allocated for the 12 p.m. flight. That news busted my mood since I was planning to visit the Eiffel Tower if I were given a 2 p.m. flight. The spare time of around 7 hours would have been adequate enough to visit Eifel and still have enough time to reach the airport a couple of hours before the flight.

I asked my cousin who resided there if I could still visit the Eiffel in this limited time window, but he advised me against this plan since the roads of Paris were accustomed to extreme levels of traffic in the morning. Like a sheep in a pack of wolves, I contemplated whether going against my cousin's advice would be a good decision. However, I ultimately decided not to delay our check-in at the airport since it would take time, and we could not afford another delay in our pursuit of landing our feet in the United States.

We did everything right from our end to make sure our plans stayed on track, but fate has its own ways of having fun. When we reached the airport, we experienced a face-palm moment when we found out that our flight was, in fact, scheduled for 2 p.m. I was frustrated, to say the least, due to the bad customer service that misguided us. Our dreams of seeing the Eiffel Tower would have come true if their customer care representative had been a little more vigilant. However, we, as passengers, were at the peril of others, so I put it behind me and planned for the journey ahead.

In the midst of engine failures and the lapse of customer service in Paris, we were eager to get on board for the United States. There is one thing I learned from this experience,

"No journey is straightforward. Even if you take every precaution and follow every manual, then again, there could be a hurdle lying far away, smiling at your fate."

The experience left its imprint on our lives, though. My wife was flying internationally for the first time and was not prepared for the delay and happenings outside the itinerary. However, the remarkable gesture of my wife providing sandwiches to the passengers who were confined at the Paris airport was heartwarming and literal proof of how human empathy can have no limits and boundaries. Life is all about perspectives. The hurdles that I faced opened my eyes to a newer set of rewards that I got. First of all, we checked in to Paris. Strictly speaking, from my wife's perspective, a person who has not traveled before is definitely going to love multiple places being introduced to him/her.

Earlier, when the plane's engine failed midway, a slew of thoughts swarmed through my mind. This time, I had to be mindful of the risks that would arise if I were to take the flight again on the same plane. The thoughts of my loved ones flashed through my mind. Their affectionate and lovely voices echoed in my ears throughout the journey. However, as the journey reached its culmination, my thoughts pointed toward all but one place: the United States. Logistical challenges eventually enhanced our problem-solving skills. My wife and I learned to pick and choose the best alternative out of a pack of adversaries. A glimmer of hope is what accompanied those whom I traveled with my life on the return to America, but surprisingly, the journey was extremely comfy and relaxing otherwise.

Fast forward a few hours, and we finally landed in New York in the first week of September 2012. As I touched down the runway, a sensation of relief sprinkled over me. The ordeal of traveling from Mumbai to the U.S. was harrowing, dampened by an engine failure, and ultimately force-confining us to Paris for one day. But it was all worth it. The sight of America made me feel as if I were turning back the clock, coming over here again for the first time. The allure of new experiences was what I anticipated throughout my flight from Paris.

I booked an apartment in Milford, Connecticut, for our stay solely because of its location. It was situated in an area where the

Indian community was abundant, and I wanted my wife to feel at home during her first experience traveling abroad. However, there was a rule that you had to wait for three weeks in order to be allowed to shift into that apartment. For that, I took help from a friend who offered me his car to travel to places in the city. I booked a hotel near my office in Connecticut to ensure that my wife was not alone for a large part of the day.

Funny as it seems, it was a consistent effort to stay as close to my wife as possible. She would get scared of the new place sometimes, so I had to ensure that I was there whenever she needed me. After three weeks of waiting time, we finally moved into our new apartment. New York was a 90-minute drive away from where I lived.

An envious yet exemplary quality of humans is to adapt to whatever they are thrown at. My wife was blessed enough to have an uncle who lived two and a half hours' drive away from our apartment. Her uncle and his wife helped my wife set up the apartment, and they were like angels to us since their help gradually made my wife feel at home.

As Asians, we are generally very touchy regarding our religion. One day, I took my wife to temples just like we had those back in India. She was overjoyed to see the multicultural aspect of American society and certainly very pleasantly happy to see how welcoming Americans are toward people from different backgrounds.

We also needed some time to adjust to different weather conditions. Spring brought along a different set of atmospheric changes. In India, we were not used to much of the snowfall, even during the last quarter of the year. However, in America, snowfall and rainfall were both very persistent at this time of the year. In October 2012, shortly after moving to the new apartment, I took the opportunity to take my wife to let her transcend into the majestic view of Niagara Falls. The place was about to be closed soon since it was closed to visitors during the wintertime every year.

Then, the month of October brought along chilly nights, warm blankets, and temperatures below zero degrees in the United States. As expected, the acute share of winds creeping and thwarting our faces near the water made us feel so cold.

I, for once, was unflustered since I had already been to the United States on several occasions, and it was easy to adapt to the treacherous weather. But my wife was the one who faced the side effects of the chilly weather, given that the temperature back in India would never be this low.

To prepare against the odds, I bought warm fleece and insisted my wife pack herself so that no passing wind entered her clothes and affected her body. As the famous saying goes:

"No matter how hard I try, it all goes bust."

The weather and its harshness tore through our precautions, and I could sense by looking at my wife's face that she was getting troubled by the cold weather. It was effectively easy to command someone to "get accustomed to the weather," but in reality, it was not as easy since our immune system runs in accordance with the data (the atmosphere) it has been fed since birth.

On Sunday morning, when the sun was out, luckily, I decided to take the ferry tour alongside my wife. The ferry tour gave my wife a clear view of the fall and made her make jaw-dropping expressions due to the exquisite scenic view.

We didn't stay there the whole day, and before night, we were back home. It was such a memorable event for my wife that she still vividly remembers every miniature detail of that day. Of the days we spent on our return to the United States, we were often under the constant fear of hurricanes, which happened during the month of November. On the contrary, there was a hailstorm in February, which my wife did not welcome as much. The roads were blocked in those days, and even taking a walk outside for groceries was impossible due to nature playing its tricks.

Does the name "NEMO" ring a bell?

It may sound like a cute name for maybe someone's adopted cat, but in our case, it was a terrifying storm named "Nemo," which was common in New York, New Jersey, Massachusetts, and Boston. In Milford, it came at full force, and around 38 inches of snowstorm were recorded on a scale. My wife was determined to see the news and was shocked to know that this Nemo storm was the biggest snowstorm in the last eighty years. It snowed for one and a half days, packing the streets with a white sheet of snow, and almost every activity in the area was paralyzed for many days.

In the aftermath of returning to America, one thing was certain: if you are not ready for calamities, then nature may hit you with force unparalleled or unprecedented. My wife's experience, overall, was excellent on her first trip to the United

States. She adapted with time and learned the art of survival in harsh weather conditions.

***

Some of the memorable moments from my first year of married life include a trip to Vegas, the Grand Canyon, visits to New York and Boston downtowns, and Saturday night boat rides on the Hudson River. Days passed by very interestingly. Occasionally, I would come home early from the office, and my wife and I would go for river walks, visit nearby parks, and enjoy nature. We explored a lot of nearby places. After a few months, my parents visited the US from India. Their original itinerary was to stay for five months, but due to some unforeseen situations, it was cut short to 3 months.

After my parents came to the US, they had a nice time with me and my wife. We visited Times Square and the Empire State Building and did some sightseeing in and around Connecticut, downtown Boston, etc. My mom was happy to visit her cousins' families. They liked the beach or shore closer to our home. Two months into their stay in the US, I received bad news from the immigration office about the denial of my visa renewal with the company. My employer gave me two weeks of grace time that the country allows as a courtesy. So, we had to pack everything and return to India within two weeks.

Since my wife was on a dependent visa, her visa, too, was out of status. Surprisingly, my parents, who were sponsored by me, were allowed to stay in the US as their visas were still valid. I hadn't really shown my parents many places yet, as they had only arrived two months before. So, we planned to show them

Niagara Falls and then drop my parents off at my sister's home in Cleveland, where she and my brother-in-law lived. The tricky part here is visiting Niagara Falls by driving ourselves. There are many cases where people, while visiting Niagara Falls, accidentally took the wrong route and ended up in Canada. Since the Falls area is so close to the border between the US and Canada, it's really easy to make that mistake. Even some of my master's classmates visited a few years back and were accidentally routed to Canada, struggling almost daily with interrogations from border security, and then came back. My manager specifically called me and informed me not to travel anywhere far during the two weeks grace time as my visa status was no longer valid. He especially warned me not to visit Niagara Falls, as there was a recent scenario where a TCS employee accidentally landed in Canada without a valid US visa stamp. The company had to spend a lot of money on immigration penalties to rectify the employee's mistake. I knew it was risky to visit the fall area, but it was one of the best places in the US for first-time visitors. So, I decided to take some risk, with the utmost caution, so as not to accidentally cross into Canada.

During the drive, just before Niagara State Park, you see boards where the road divides; one goes to New York's Falls State Park, and the other towards the Rainbow Bridge, which leads to Canada. I stopped at the shoulder a couple of times to double-check and confirm, as I didn't want to take any chances with my invalid immigration status. To your surprise, do you know what actually happened? Don't worry; this time, it wasn't our turn to be surprised. We safely reached the falls, and my parents had a really nice time with the tours at the falls. After that, we drove to

Cleveland to my sister's place, dropped my parents off, and my wife and I returned back to Connecticut. We had another five days to wind up, pack up all our stuff, sell all our leftover furniture, and head back to India. The first time I came to India, I flew on Emirates Airlines, and I used to praise it a lot to my wife. So, she asked if we could go back to India on that airline. I asked our office's travel department if we could get the desired airline. Luckily, we got an Emirates itinerary and had a nice journey back to India via Dubai.

Chapter 10: The Aftermath

Life is unpredictable, and it comes with a lot of surprises for every single soul. For me, this surprise came in the form of another trip to the United States of America. Yes, US! She was waiting for me again. I remember my best friend used to say this a lot: "If we encounter hurdles, don't panic, and understand that it is a sign that nature wanted us to pause that task instead of trying hard to achieve it." That advice worked. It is important to understand that all these sayings are not implemented in every situation, and of course, they vary based on scenarios and environments, whether we speed up or slow down. It is important to understand the situation and master the formula that suits a particular situation.

This is life in its truest essence!

I lived in the United States for six years, and when I left, I thought I wouldn't be returning to the US again. I thought even if I got a chance in the future, it would only be a short business trip. I started adapting to stay back in my home country, but career growth-wise, I did not get a breakthrough. Thus, I decided to ask my company for a transfer from Chennai to Hyderabad in India, where I bought a small flat that I thought was my destination, but an old project knocked on my door. It was a good opportunity for me. I got a call from an old project lead saying that there was an on-site opportunity, and I availed myself of it. I traveled to the UK in 2014, and my visa was for a year. The original plan was also to stay in the UK for one year, but due to a bad work-life balance in that project, I decided to move out within eight months.

During that hectic schedule, a friend from the US referred me to a company that could apply for an H1B visa for me. I initially did not want to go to the US again as I had already planned to shift to Hyderabad and lead a low-profile life, and the UK project experience is also one of the reasons for a bad work life. At a point in time, I thought of quitting IT completely and opening a small business in my hometown, but my friend convinced me to apply for H1B, which is a lottery system. That year, approximately three hundred thousand applications were submitted, and only sixty-five thousand were picked up. Pretty much one out of five chance, and after my petition gets picked up, it would undergo the process.

In some cases, the petition was rejected even after being picked up in the lottery. Even if the petition was approved, the next step was Visa stamping, which, in some rare cases, people got rejected due to missing documentation. I was unsure until all these steps were passed, but as I said, life had a lot of unknown surprises for us. My petition got picked up in the lottery, and my wife and I got visas stamped in the Hyderabad embassy.

I landed again in the US on 31st December 2014, just after a sixteen-month break. This time, I traveled alone first, got a job, found an apartment, and then brought my wife to the US again. All happened as we expected in a decent amount of time. This time, my first job was in Los Angeles. We were there in LA for a short time but had a lot of fun due to weather conditions. Since this job was a three-month extension contract each time, I had to wait until the end of these three months to know whether I'd get an extension or not. Hence, I decided to search for a long-term opportunity. During that course, we landed in Cincinnati,

OHIO, in 2015. From that time on, Cincinnati became our hometown, which was far from the ocean, unlike all my previous places. Later, we were blessed with twins, a girl and a boy, in 2019. My kids brought me a stroke of triple luck. I changed jobs one more time with a hike and got a team that respects work-life balance to a great extent. I believe that productivity is at its peak when you have the best team to work with. We also had a good friend group in Cincinnati, which was one of the primary reasons we stuck to this city.

I always wanted to take my wife to Florida, especially Tampa, where I landed first in the US, lived a couple of years before marriage, and show her the beaches there. In the US, especially when you are on the East Coast, a Florida trip is not a big deal. We booked flights a couple of times in 2016 and 2017, but both were canceled at the last minute due to hurricanes. I'm unsure if we hindered the flight journey, beaches, or Florida.

To test that, I booked one more time in 2020. Usually, I planned for a trip just a month before, but this time, I planned the journey almost three months before with a full itinerary. We wanted to take our 1-year-old twins also with us to Miami. The travel plan was in April 2020, and as you all know, from March mid-onwards, the entire world shut down due to COVID-19. I don't want to use or rotate around that word, but it's a situation to tie into. Whatever the case, our third attempt was a failure, too. There will be billions of people impacted by 2020, and many of us were part of it. We were here to take only good by learning from the bad.

I want to disclose another mini adventure flight trip with some delays from the airlines and how we went through them with kids. On our previous five-day trip from India to the US, it was just me and my wife. We had a twin child with us this time, which was challenging. My brother-in-law's marriage got fixed, and we planned to visit India. My wife wanted to go a few months before the wedding to help her parents and her brother shop. I could not take that long a vacation since the kids were too young. I decided to drop my wife and kids off and come back in a week, travel again for the wedding time, and bring my family back. The itinerary was to and fro from Chicago to Delhi and from Delhi to Vizag.

Since we lived in Cincinnati, we drove to Chicago and stayed at my wife's uncle's home. They were very nice in accommodating us and drove us to the airport several times. Are you wondering why I said it several times? To our surprise, we got delayed a day on both sides. It happened when I dropped my family and brought them back. While going from Chicago, the flight was at midnight. After the check-in process, they mentioned the flight was delayed for seven hours as the inbound flight had not arrived, and they asked us to come back in the evening. My wife's uncle came back to pick us up and dropped us off again in the evening.

The evening, after an hour of dilemma, we found the airline staff and learned that the flight had been canceled for the day and we had to come back the next day. My wife and I started recollecting our previous experiences and got scared, but nothing could be done. So, we asked our relative to come again to pick us up, and we decided to go by cab the next day, but my wife's uncle was so patient and dropped us off. The next day, the flight took off finally. Later, we got to know that the previous day's flight did not arrive because it had to be routed to a different country due to a heart attack for a passenger on the flight.

Since the first flight got delayed for a day, none of the passengers could catch their connecting domestic flights from Delhi to various cities in India. Since the original delay was due to airlines, we expected they would arrange a connecting flight for all the passengers. We landed in India around 1 pm local time, took an hour to finish the entry immigration checks and get baggage, and the actual twist started. There is a gated transfer for all connecting flights, and all the passengers with baggage piled up there, waiting for airline officials to address them.

Multiple airline professionals try to talk to customers, but none of them assure them of anything. Passengers started shouting at the officials to provide an immediate plan for their connecting flight. We know our connecting flight has gone that morning, and no more flights until the next day, but some of the connecting flights are available that day evening, so those passengers are curious to know what airlines are going to do. It's a challenge for airlines to accommodate new passengers at the last minute due to seat availability issues. But the surprising part is that they did not even plan what to say to customers or accommodate customers until all the passengers arrived.

After a two-hour confusion there, one of the porters suggested that there is a back office of this airline at Gate 1. There is no seating space there, but you can inquire there, and they may help. Knowing there was no seating space, we still decided to go there as I had my family and baggage. I cannot leave my family and go alone as we don't have a local mobile with us, also not sure if airport WIFI works properly, and it will be hard for my wife to manage both the kids. Once we reach there, we find a huge crowd in front of the counter, no proper queue, and everyone is on top of the counter person to get their task done quickly.

Beauty is none of the airline staff volunteering to ask people to stand in line, and they are going to accommodate everyone, nor saying something like passengers can go to a hotel, and airlines can send the passes based on availability. My wife sat on the floor with both of our infant kids sleeping on her lap, two baggage strollers next to her. Luckily, we bought some food and water while walking to gate one back office. I was standing in that crowded line for four hours. After that, the counter person took my international flight boarding pass and mapped it for the next day's flight. He printed the next day's flight passes for my wife and kids and said I would get my boarding pass to the hotel where I stayed. It was hard to believe whether they could send the boarding pass to the hotel later, but I had no other option.

It was already 7 p.m. I was too tired to stand further and wanted to go to the hotel as soon as possible. There was another counter next to that for issuing hotels to the passengers. That person at the counter had a rule that we had to show our connecting domestic flight boarding pass, which was newly issued to her, and only then would she issue a hotel room for us. You know, that hotel issuing process was like a fifty-year-old back process. There was a ruled notebook, and she wrote each passenger's name, date of birth, passport number, itinerary details, US address, local address, contact phone numbers, etc. She was taking her own time to write each passenger's name in between taking a five-to-ten-minute break to call the hotel, book the room, and get their confirmation. There were multiple staff members, and the managers were coming and going anxiously, but none of them took time to answer passengers or quicken the process. After a one-hour wait at the hotel issue counter, I thought of booking a hotel myself and going, but I had to stay in an airline-provided hotel. Only then did I get my boarding pass to the room. So, it was pretty much hands-tied with no escape.

During those five to six hours, I connected to the airport's wifi and tried all other options, calling my cousins and searching for other airlines' domestic flights for the next one to two days, and it seemed everything was booked. They issued a hotel after a two-hour extended marathon at the second counter. Another person came and took a bunch of passengers along with us to the airport shuttle area. A couple of single passengers offered us space to get into a cab.

Around 9 pm, we started our trip and reached the hotel in twenty minutes. After a twenty-minute wait in the lobby, we finally landed in the room around 10 p.m. Due to the jetlag, the kids slept for hours and woke up after we reached the hotel room. After a sixteen-hour flight and standing for eight hours at the airport, the hotel room looked like heaven. We took the next day's flight and reached our native place. I only dropped my family and returned to the US in a week.

I went to India to attend a wedding after a couple of months. It was my brother-in-law's wedding. Ten days before my wife's brother's wedding, we had a nice time visiting our parents' and siblings' places and a few new restaurants in Vizag. We attended the wedding ceremony and planned to return to the US with family in February 2022. Due to COVID restrictions at that time, all of the passengers were required to take a COVID test, which was less than twenty-four hours of the travel time, and can travel only if it is a negative result.

As those rules may change frequently, I called to check with airlines regarding the latest rules, and they said younger children are not required to show any proof of a COVID test. So, my wife and I took the test, and luckily, we got a report within twelve hours. This was again a connecting flight from Vizag to Delhi and Delhi to Chicago. This journey was again an unforgettable memory for us. At the check-in counter in Vizag, they asked for the negative COVID-19 report for all family members. I informed them that, as per the latest rules, I confirmed with airline customer care that younger kids were not required to show any proof. But the person at the counter said there is no such rule.

When I asked why the customer care people mentioned that, they doubted me, saying, "You would have called some customer care and not the airline's customer care." I got angry but controlled myself as I wanted to somehow finish this trip with the kids smoothly.

After a bunch of discussions, they offered an alternative, as the requirement was for international flights and not for domestic flights, so they could allow travel from Vizag to Delhi. While boarding in Delhi, we had to produce test reports for all four of us. They have issued boarding passes for domestic flights and asked us to collect the international flight boarding passes in Delhi. Since we have a couple more hours in Vizag and a 5+ hour layover in Delhi, I took that option, thinking we would get the boarding passes for our international flight in Delhi. We managed to find a testing center that offered mobile sample pick-up and expedited results. The technician visited the airport in less than an hour to take our kids' samples for the COVID test. They said they could send the online copies of the results within three to four hours, which was still within our time frame.

By the time we landed in Delhi, the reports were available in my email. For international transfer, we had to pick up our baggage, which consisted of eight check-in bags along with two cabins, and check in again. We picked up our baggage and went to the international check-in counters, but we still had 4 hours and 30 minutes before departure time. We were happy that we made it on time and had a decent amount of time to finish all the procedures. Since we came pretty early, we stood for about fifteen minutes in line and were able to approach a boarding staff. The counter person checked our names from the passports and asked us to go to another counter on the other side, and behind these counters, we had to go roundabout to go to the other side. We didn't know that there were two different sets of counters.

Before going, I asked what this line was, what the other line was, and why we had to go there. The person mentioned that you have to go there for your itinerary. I didn't want to waste any more time, so we visited the other counter and moved all of our carts and strollers. It was eight check-ins, two cabin baggage, two luggage carts, and two kid strollers. We still had more than four hours to depart. I gave the four passports and four test reports to the person at the counter. He gave back everything and asked to give only passports, and he would ask for the remaining ones one by one. I did the same. He asked me to confirm my first and last name and type one letter at a time. He saw my passport but still wanted me to spell each letter, and he was typing damn slow.

I remembered the DMV seen in the Zootopia movie, but I couldn't help.

After five minutes, he said he found my record. I asked if he could issue boarding passes, and he said he had to check a few things. After reviewing the passports for 5 minutes, he said that I had only two months left on my Visa stamping validity. I said I already had another approved petition that was good for three years after these two months. Hence, I was still eligible to travel. He was trying to ask more questions on my immigration status, etc., and then I reminded him that those questions were supposed to be handled at the immigration check before going to the gate and asked why he was worried here when I had a valid passport and Visa. He then moved on and asked why he did not get boarding passes in Vizag.

When I explained our situation to the airline representative, he requested our test reports. Handing them over, I watched as he took another five to ten minutes to validate them. The tension grew as he scrutinized each of us, meticulously comparing our faces to the passport photos, even those of our young children. It felt like he was deliberately dragging his feet, wasting our precious time. I pointed this out with a line of passengers growing behind us, only to be dismissed with a comment about other available counters.

Frustrated and anxious, I pressed him on why there was a delay in issuing our boarding passes. He claimed there was an issue with seat availability and that he was consulting with someone to resolve it. Desperate, I suggested that if four seats weren't available, he could at least provide two pairs so each could look after one child. He assured me he would try, but his lack of urgency was palpable.

As the minutes ticked by—nearly forty-five in total—my frustration turned to disbelief when he finally informed us that the flight was overbooked. We would have to travel the next day. I was stunned—how could this be happening when we had booked our tickets five months in advance? The airline's website had failed us repeatedly, redirecting us to error pages and preventing us from pre-booking seats. Now, we were at the mercy of the airport counter, and it had let us down spectacularly. The representative offered no solutions, only directing me to speak with his manager if I had further questions.

Leaving my wife and children with our baggage, I hurried to another line to speak with the manager, my frustration mounting with each step. After a twenty-minute wait, I finally reached her. She presented two equally disheartening options: wait and travel the next day or receive a refund for the one-way airfare and try to book another airline. Given the unlikelihood of finding last-minute international tickets, the second option was a non-starter.

I couldn't contain my frustration. How could the airline overbook and leave a family with young children stranded? She insisted we should have been at the airport four hours before departure, implying our tardiness was to blame. I explained that we had arrived almost 4.5 hours early, but my words fell on deaf ears. The airline's incompetence was glaringly apparent, each step of the process highlighting their disregard for our situation. As I checked online for other flights, I found nothing available in the next few hours, and a sense of helplessness washed over me.

After exploring our options, I discovered that there was another flight available in ten hours. However, the prices were exorbitantly high—almost four to five times the cost of our current tickets. Given the choice of either waiting for ten hours and paying a steep price or waiting an additional day to travel with our current airline, I opted for the latter. I asked the manager if the airline could provide hotel accommodation along with next-day flight tickets. She confirmed that they could, but there was a process involved. Additionally, the airline offered to hold our baggage for the day, which was a significant relief.

I was already familiar with the process of securing hotel accommodation due to a similar delay during our inbound travel, where we stayed in a hotel overnight in Delhi. Like us, other passengers affected by the overbooking were anxious about how to get hotel accommodation and wanted to stay together. They turned to the airline for guidance. I shared my previous experience, explaining that we needed to go to Gate 1 at the back of the office to get boarding passes, followed by another counter for hotel arrangements. Hearing this, the other passengers smiled, relieved to have some direction. A couple of them asked why I was still traveling with this airline, given our previous delay. I explained that this was a round-trip itinerary and I hadn't anticipated delays on both ends.

With our new plan in place, we proceeded to the back-office counter in a more organized manner, now fewer in number but united in purpose. Some passengers hoped the process would be quick and expected to reach the hotel within an hour. I warned them that the procedure was manual and time-consuming and that they should be prepared to spend at least four to five hours.

Initially shocked, they soon adjusted their expectations. After a three-to-four-hour wait, we finally boarded a shuttle to the hotel, arriving at our rooms around midnight.

The silver lining was that we didn't have to carry our check-in baggage. We spent the entire next day at the hotel in Delhi, hesitant to venture out with an upcoming international flight pending. Instead, we focused on relaxing and preparing for the fifteen-hour journey ahead. Reflecting on the situation, I couldn't help but think that the airline should have informed us about the overbooking issue while we were still in Vizag. That way, we could have spent an extra day with our parents instead of being stuck in this limbo.

We arrived at the airport nearly six hours before our flight the next day, determined to ensure our departure. We waited for the check-in counter to open, weighed our cabin baggage, received a new set of boarding passes, and prepared for our journey. While at the check-in counter, one of the porters recognized us from the previous day. I explained our predicament, and he pointed to the counter across from us, explaining that those passengers had also been removed from their flight due to overbooking. He revealed that this was a regular practice for the airline—bumping passengers to the next day's flight in a continuous cycle.

Curious and frustrated, I asked the check-in attendant how the airline decided which passengers to remove due to overbooking. He replied that it was a random process with no strict rules in place. This lack of consideration for families with infants or elderly passengers was disheartening. Reflecting on the situation, I realized the underlying reason for the overbooking. Our travel was on a weekend, a peak time when many people are willing to pay exorbitant fares at the last minute. The airline capitalized on this by removing passengers and rebooking them on later flights, continuing this practice until mid-week when there were typically enough vacancies to accommodate everyone.

This practice was not unique to our flight. In general, weekend demand is higher than weekdays, driving up prices. When booking airfares, it's common to find cheaper rates mid-week, particularly on Tuesdays and Wednesdays, when flights are less likely to be fully booked. The airline exploited these vacancies by shifting passengers from overbooked weekend flights, maximizing their profits at the expense of passenger convenience.

For instance, the airline might spend $100 on a hotel room for a displaced family but could gain $1,000 to $1,500 per ticket by selling overbooked seats at the last minute. Even if they compensated displaced passengers with hotel stays for three or four days, the net gain was significant—potentially $5,000 to $6,000 for a few tickets. While profitable for the airline, this strategy was a deeply unfair way to generate revenue, causing considerable distress to passengers.

Though we arrived six hours before our departure, the situation was far from favorable, though this time, it wasn't the airline's fault. At the last minute, we received a call from an elder relative asking us to accompany their daughter, who was traveling on the same flight to the US with her six-month-old baby. Given that our twins were also under three years old and had already kept us busy, it was a challenging request. However, we didn't see a compelling reason to refuse. After checking in our baggage and obtaining our boarding passes, we thought we had ample time—over four hours—to complete the immigration and security checks. However, the relative's daughter arrived late, causing a delay. That hour of waiting felt like an eternity, especially since we had our boarding passes and were ready to proceed. Wanting to avoid any tension with my wife, I waited patiently.

When she finally arrived with her baggage and infant, we encountered another issue: her baggage was overweight, and the airline wouldn't allow it. We had to rearrange our luggage, utilizing our two lightly packed cabin bags and an extra bag. After these adjustments, she received her boarding pass, and we headed to the immigration check. We hurried through the process with only 90 minutes left until departure, but the long lines ate up our time. With just 25 minutes left, we sprinted toward the gate, which was far from the security check area. Balancing two strollers, two cabin bags, and the relative's baby, along with her non-wheeled cabin bag, made the dash even more hectic. Just when it seemed we might miss the flight, a godsend in the form of a battery-powered cart appeared. Without that ride, we would have certainly missed our flight. Miraculously, we boarded just seconds before the gate closed.

This journey, now over two years ago, was followed by a much smoother domestic trip. As we await our next international adventure, I can't help but reflect on the unpredictability and chaos that often accompany travel. Despite the challenges, these experiences shape our resilience and adaptability. Here's to hoping for less drama in our future travels, and I would appreciate your well wishes for our next adventure ☺.